Optimisation

Longman Mathematical Texts

Edited by Alan Jeffrey

Elementary mathematical analysis I. T. Adamson
Elasticity R. J. Atkin and N. Fox
Set theory and abstract algebra T. S. Blyth
The theory of ordinary differential equations J. C. Burkill
Random variables L. E. Clarke
Electricity C. A. Coulson and T. J. M. Boyd
Waves C. A. Coulson and A. Jeffrey
General relativity J. Foster and J. D. Nightingale
Optimisation D. M. Greig
Integral equations B. L. Moiseiwitsch
Functions of a complex variable E. G. Phillips
Special functions of mathematical physics and chemistry I. N. Sneddon
Continuum mechanics A. J. M. Spencer

Longman Mathematical Texts

Optimisation

D. M. Greig

Department of Mathematics, University of Durham

Longman
London and New York

Longman Group Limited London

Associated companies, branches and representatives throughout the world

Published in the United States of America by Longman Inc., New York

First published 1980

British Library Cataloguing in Publication Data

Greig, D M
Optimisation. – (Longman mathematical texts).
1. Mathematical optimization
I. Title
515 QA402.5 79-42892

ISBN 0-582-44186-2

Printed in Great Britain by McCorquodale (Newton) Ltd Lancashire.

Contents

Introduction ix

Notation xii

1: Basic results 1

1.1 Classical theory 1
1.1.1 Definitions 1
1.1.2 Conditions for unconstrained variables 1
1.1.3 Equality constraints 6
1.2 Inequality constraints 12
1.2.1 General considerations 12
1.2.2 Necessary conditions, Kuhn–Tucker form 13
1.2.3 Convexity, Farkas' Lemma 15
1.2.4 Abnormal point 18
1.2.5 Sufficient conditions 19
1.3 Duality 22
1.3.1 Saddle-point conditions and duality 23
1.4 Summary of basic results 28
Notes 29
Problems 31

2: Unconstrained optimisation 35

2.1 Line-search methods 35
2.1.1 Dissection methods 35
2.1.2 Fitted polynomial methods 38
2.2 General search methods 40
2.3 Gradient methods 41
2.3.1 Steepest descents 42
2.3.2 Gradient properties of quadratic functions 44
2.3.3 The method of conjugate gradients 46
2.3.4 Powell's method 49

2.4 Newton and quasi Newton methods 54
2.4.1 Newton's method 54
2.4.2 Davidon–Fletcher–Powell method 56
2.4.3 General considerations on matrix updating methods: the Huang family 59
2.4.4 Common directions for the Huang family 61
2.4.5 The conjugate-gradient algorithm revisited 62
2.4.6 Other methods 63
2.5 Summary of methods 64
Note 64
Problems 65

3: Linear programming 70

3.1 Solution of LP problems 70
3.1.1 Statement of the problem 70
3.1.2 Theory of linear problems 71
3.1.3 Simplex method 77
3.1.4 Starting the simplex method 81
3.1.5 Revised simplex method 83
3.1.6 Degeneracy 86
3.2 Duality 86
3.2.1 Duality theorem 86
3.2.2 Lagrange multipliers 90
3.2.3 Dual simplex method 91
3.2.4 Sensitivity and duality 91
Problems 94

4: Applications of linear programming 98

4.1 Allocation of resources 98
4.2 Transportation problems 100
4.2.1 Definitions, general properties 100
4.2.2 Transportation algorithm 102
4.2.3 Transshipment problems 106
4.2.4 Assignment problems 109
4.2.5 Assignment algorithm 111
4.3 Game theory 116
4.3.1 Definition of problem 117

4.3.2 Determination of optimal strategies 119
4.3.3 Further aspects of game theory 120
4.4 Separable programming 122
Note 125
Problems 125

5: Constrained optimisation 132

5.1 General properties of the solution 133
5.2 Projection methods 134
5.2.1 Some projections 134
5.2.2 Method of solution of inequality-constrained problems using projection 139
5.2.3 Projection methods – examples 142
5.3 Quadratic programming 145
5.3.1 Modified simplex method 145
5.3.2 Projected Newton method 146
5.4 Application of projection methods to nonlinear constraints 147
5.5 Penalty function and multiplier methods 150
5.5.1 Sequential unconstrained minimisation 150
5.5.2 Lagrangian methods 156
5.6 Summary of methods of constrained optimisation 163
Problems 163

Hints and answers to problems 168

References 175

Index 178

Introduction

The problem considered in this book is that of optimising (i.e. maximising or minimising) the numerical value of a function of one or more variables which may be subject to constraints. The requirement is to

$$\text{minimise } f(\mathbf{x})$$

where f is the objective function and $\mathbf{x}=(x_1, x_2, \ldots, x_n)$ is an n-dimensional vector, subject to

$$c_i(\mathbf{x})>0, \quad i=1,2,\ldots,k$$

$$c_i(\mathbf{x})=0, \quad i=k+1,\ldots,m.$$

There is no loss of generality in taking a minimising problem only, since maximising f is the same as minimising $-f$, and similarly the inequality constraints may all be written one way. If $m=0$ the problem is unconstrained.

This mathematical formulation is common to problems arising in widely different situations and the theory of optimisation methods provides a unified approach to their solutions. Physicists, chemists and engineers have been interested in design optimisation, for example the design of a chemical plant to maximise production subject to resource and technical limitations and known relationships between conditions and output. Scientists have used nonlinear curve fitting and nonlinear statistical models whose implementation requires minimising sums of squares or maximising a likelihood function. Economists, operations research analysts and planners have considered the best allocation of limited resources in industrial, social and military situations. These last are mostly linear models and the corresponding optimisation technique, known as linear programming, is the most widely used of all. There are already many other applications and more are continually being developed.

There is then no doubt that there is a widespread practical

requirement to solve optimisation problems. There is also an elegant and comprehensive theory underlying the approach to their solution, and it is the aim of this book to give some account of this. Only problems which have the above structure are considered, and in addition there are required a few desirable extra properties. The vector $\mathbf{x}$ is considered only at one point in time so that the problem is static, and the relation is deterministic, that is the value of f is determined by the value of $\mathbf{x}$. There is no discussion of control theory, or of dynamic programming, or of the optimisation of functionals, as in problems of calculus of variation type. The function f is assumed continuous and in some sense smooth and $\mathbf{x}$ is subject only to the constraints shown and is not, for example, restricted to have integer values. Within these limitations the book seeks to present the theory underlying current methods of solution of the optimisation problem. While it is not possible to go into details of the various algorithms or of the computer programs which implement them, it is hoped that sufficient material is included to show the interrelations of methods suggested for different conditions.

In Chapter 1 the classical theory of optimisation with equality constraints is presented using Lagrange multipliers. The corresponding results in the inequality-constrained case, the Kuhn–Tucker conditions, are then described with a discussion of convexity. The elegant theory of duality inherent in such problems is introduced.

Chapter 2 describes methods for finding a minimum in the unconstrained case. These are all iterative, proceeding from an initial point $\mathbf{x}$ by successive improvements, but differ in the way these improvements are made. Search methods, gradient methods and quasi-Newton methods are described – these labels are used for convenience though many algorithms might be listed under more than one heading.

Chapter 3 deals with the special case of linear programming, that is constrained optimisation in which the function f and the constraints c_i are all linear in $\mathbf{x}$. This is important theoretically because of its simple structure and practically since there exists an efficient algorithm, the simplex method, guaranteed to yield the solution in a finite number of iterations. This and its variant, the revised simplex method, are described and illustrated, and shown to solve the dual problem also.

Chapter 4 deals with applications and special cases of linear programming. Game theory and sensitivity are shown to be closely linked with the dual problem; and special algorithms are described for transportation, trans-shipment and assignment problems.

Chapter 5 describes methods for nonlinear programming. Particular cases are quadratic programming – f quadratic, c_i linear – and convex programming – f convex, c_i concave – for which a local minimum can be proved to be a global minimum, a property which is not true generally. Methods used to solve constrained problems are of two general types, depending on their treatment of the constraints. In penalty-function methods the constraints are incorporated into a modified objective function which is then treated as unconstrained. Augmented Lagrangian methods are of this type. The other approach is to use the constraints to guide the sequence of values of $\mathbf{x}$, and this gives rise to projective and Newton-type methods

A number of problems are included at the end of each chapter. The numerical ones are very small scale and included simply to illustrate the methods. The student will find package programs implementing the main algorithms in any computer centre.

Notation

The following notation will be used consistently throughout this book.

$\mathbf{x}, \mathbf{x}_r$	any boldface lower case symbol is a column vector.
x_i, x_{ri}	the i-th components of $\mathbf{x}, \mathbf{x}_r$ respectively
$\mathbf{x}^{\mathrm{T}}$	the transpose of $\mathbf{x}$, that is a row vector.
$f(\mathbf{x})$	a scalar function of $\mathbf{x}$; when $\mathbf{x} \in R^n$ this can be written $f(x_1, x_2, \ldots, x_n)$
$\mathbf{c}(\mathbf{x})$	a column vector each of whose components c_i is a function of $\mathbf{x}$
$\mathbf{g}(\mathbf{x})$	the gradient $\nabla f(\mathbf{x})$, that is the column vector such that $g_i(\mathbf{x}) = \partial f/\partial x_i$
$\mathbf{a}_i(\mathbf{x})$	the gradient $\nabla c_i(\mathbf{x})$
$\mathbf{A}(\mathbf{x})$	the matrix whose rows are $\mathbf{a}_i^{\mathrm{T}}(\mathbf{x})$
$\mathbf{a}^j(\mathbf{x})$	the j-th column of $\mathbf{A}(\mathbf{x})$
$\mathbf{G}(\mathbf{x})$	the Hessian of $f(\mathbf{x})$, that is the symmetric matrix whose elements are $G_{ij} = \partial^2 f/\partial x_i\, \partial x_j$
X	the feasible set: $X = \{\mathbf{x} \mid c_i(\mathbf{x}) \geqslant 0, i = 1, 2, \ldots, m\}$
I_r	the locally active set of constraints at $\mathbf{x}_r$; $I_r = \{i \mid c_i(\mathbf{x}_r) = 0\}$
$\mathbf{C}_i(\mathbf{x})$	the Hessian of $c_i(\mathbf{x})$
$L(\mathbf{x}, \boldsymbol{\lambda})$	the Lagrangian, $f(\mathbf{x}) - \sum \lambda_i c_i(\mathbf{x})$
$\mathbf{F}(\mathbf{x}, \boldsymbol{\lambda})$	the Hessian of the Lagrangian, $\mathbf{G}(\mathbf{x}) - \sum \lambda_i \mathbf{C}_i(\mathbf{x})$
$\mathbf{x}^*, f^*$	the minimum point and minimum value respectively

1 Basic results

1.1 Classical theory

1.1.1 Definitions

Given a function $f(\mathbf{x})$ of an n-dimensional variable $\mathbf{x}$ and a region X, possibly the whole space R^n, within which $\mathbf{x}$ must lie, then a minimum of f can be defined in the obvious way.

(1) $f(\mathbf{x})$ has a *strong local minimum* at $\mathbf{x}^*$ if $\mathbf{x}^*$ is in X and there exists a neighbourhood of $\mathbf{x}^*$ also in X such that $f(\mathbf{x})$ is larger than $f(\mathbf{x}^*)$ at all points $\mathbf{x}$ of this neighbourhood; specifically if there exists $\varepsilon > 0$ such that $f(\mathbf{x}) > f(\mathbf{x}^*)$ for all $\mathbf{x} \in X$ such that $\|\mathbf{x} - \mathbf{x}^*\| < \varepsilon$. (Here $\|\mathbf{v}\|$ means the Euclidean norm $[\sum_{i=1}^{n} v_i^2]^{1/2}$.) If the relation is $f(\mathbf{x}) \geqslant f(\mathbf{x}^*)$, then $\mathbf{x}^*$ is a *weak local minimum.*

(2) $f(\mathbf{x})$ has a *global minimum* at $\mathbf{x}^*$ if $\mathbf{x}^*$ is in X and $f(\mathbf{x}) \geqslant f(\mathbf{x}^*)$ for all $\mathbf{x}$ in X.

Local conditions can be derived assuming some continuity in f. Adopt the usual definitions that f is of class C^1 for $\mathbf{x} \in X$ if it is continuous in X and has continuous first-order partial derivatives in X, and of class C^2 in X if it is C^1 and also has continuous second-order partial derivatives in X. Then we can describe the behaviour of f near $\mathbf{x}^*$ in terms of derivatives. However it is necessary to distinguish between the two cases (i) $\mathbf{x}^*$ is in the interior of X and (ii) $\mathbf{x}^*$ is on the boundary of X. Clearly if $\mathbf{x}$ is unconstrained, then (i) is always true, so we develop first conditions which apply to this case.

1.1.2 Conditions for unconstrained variables

When X is R^n, then for f of class C^1, and any $\mathbf{p}$,

$$f(\mathbf{x}^* + \varepsilon\mathbf{p}) = f(\mathbf{x}^*) + \varepsilon\mathbf{p}^{\mathrm{T}}\mathbf{g}(\mathbf{x}^*) + O(\varepsilon^2), \tag{1.1}$$

where $\mathbf{g}(\mathbf{x})$ is the gradient vector ∇f at $\mathbf{x}$

$$\{\mathbf{g}(\mathbf{x})\}^{\mathrm{T}} = \left\{\frac{\partial f}{\partial x_1}, \frac{\partial f}{\partial x_2}, \ldots, \frac{\partial f}{\partial x_n}\right\}. \tag{1.2}$$

Consider the change in the value of f due to a move of length ε in the direction of a unit vector $\mathbf{p}$.

$$f(\mathbf{x}+\varepsilon\mathbf{p}) - f(\mathbf{x}) = \varepsilon\mathbf{p}^{\mathrm{T}}\mathbf{g} + O(\varepsilon^2)$$

and so the limit as $\varepsilon \to 0$ of

$$\{f(\mathbf{x}+\varepsilon\mathbf{p}) - f(\mathbf{x})\}/\varepsilon$$

is $\mathbf{p}^{\mathrm{T}}\mathbf{g}$, and this is the directional derivative of f in the direction $\mathbf{p}$. If $\mathbf{g}=0$, the point $\mathbf{x}$ is stationary – that is, a move in any direction gives a zero rate of change of f; and if $\mathbf{g}$ is not zero, then it points in the direction of greatest rate of change of f at $\mathbf{x}$. For by the Cauchy–Schwartz inequality (Note 1.1),

$$(\mathbf{p}^{\mathrm{T}}\mathbf{g})^2 \leqslant (\mathbf{p}^{\mathrm{T}}\mathbf{p})(\mathbf{g}^{\mathrm{T}}\mathbf{g}) = \mathbf{g}^{\mathrm{T}}\mathbf{g},$$

since $\mathbf{p}$ is a unit vector. But clearly $(\mathbf{p}^{\mathrm{T}}\mathbf{g})^2 = \mathbf{g}^{\mathrm{T}}\mathbf{g}$ if $\mathbf{p}$ is in the direction of $\mathbf{g}$ or $-\mathbf{g}$; and so the greatest directional derivative is in this direction. The function values increase in the direction $+\mathbf{g}$, decrease along $-\mathbf{g}$.

From definition (1) it follows that

(3) a necessary condition for $f(\mathbf{x})$ to have a local minimum at $\mathbf{x}^*$ is

$$\mathbf{g}(\mathbf{x}^*) = 0 \tag{1.3}$$

This condition is sufficient for $f(\mathbf{x})$ to have a stationary point at $\mathbf{x}^*$; such points may also be local maxima or saddle points as illustrated by the contours in Fig. 1.1, in which f is a function of two variables x_1, x_2.

To distinguish between these cases requires at least second-order conditions. If f is C^2 then when $\mathbf{g}(\mathbf{x}^*)=0$, (1.1) can be extended to give

$$f(\mathbf{x}^*+\varepsilon\mathbf{p}) = f(\mathbf{x}^*) + \tfrac{1}{2}\varepsilon^2\mathbf{p}^{\mathrm{T}}\mathbf{G}(\mathbf{x}^*)\mathbf{p} + O(\varepsilon^3), \tag{1.4}$$

where $\mathbf{G}(\mathbf{x})$ is the symmetric Hessian matrix

$$G_{ij}(\mathbf{x}) = \frac{\partial^2 f}{\partial x_i\, \partial x_j}. \tag{1.5}$$

(4) So an additional necessary condition if $\mathbf{x}^*$ is a minimum of f

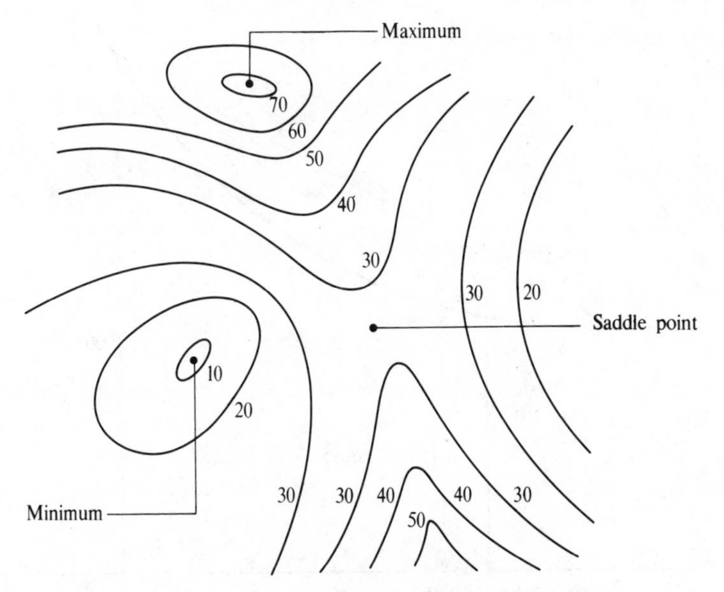

Fig. 1.1. Contours of function of two variables.

in this case is that $\mathbf{G}(\mathbf{x}^*)$ is positive semidefinite, and necessary and sufficient conditions for f of class C^2 to have a strong local minimum at $\mathbf{x}^*$ are

$$\mathbf{g}(\mathbf{x}^*)=0, \quad \mathbf{G}(\mathbf{x}^*) \text{ strictly positive definite.} \tag{1.6}$$

Note that these are local properties which give in general no guidance about global behaviour – except that, from definitions (1) and (2), the global minimum must also be a local minimum. Thus to find the global minimum, which is the quantity of greatest practical interest, it is necessary to select that one among all the local minima at which $f(\mathbf{x})$ has the smallest value. However there are certain restricted types of function f for which more can be said.

(5) If $f(\mathbf{x})$ is the quadratic function

$$f(\mathbf{x})=\tfrac{1}{2}\mathbf{x}^{\mathrm{T}}\mathbf{G}\mathbf{x}-\mathbf{b}^{\mathrm{T}}\mathbf{x} \tag{1.7}$$

where $\mathbf{G}$ is a symmetric matrix independent of $\mathbf{x}$, then

(i) $f(\mathbf{x})$ has a strong local minimum if and only if $\mathbf{G}$ is positive definite. In this case

(ii) the minimum is at $\mathbf{x}^*=\mathbf{G}^{-1}\mathbf{b}$

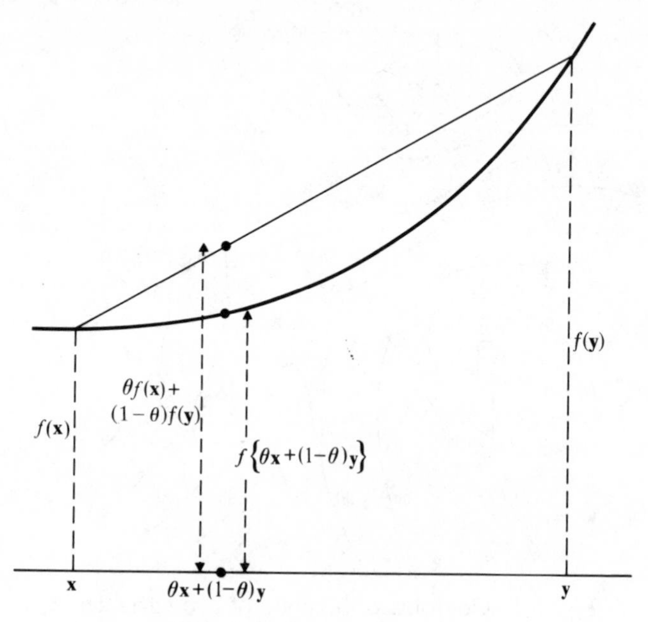

Fig. 1.2. Convex function.

(iii) the minimum point is unique and hence is the global minimum.

These results follow immediately from (1.6) and the fact that $\mathbf{G}$ must be nonsingular if it is positive definite. (Note that if $\mathbf{G}$ is positive semidefinite, then it is possible to have a strong minimum but other conditions need to be imposed – see Problem 3.)

A wider class of functions with similar properties are *convex functions* defined as follows:

(6) A function $f(\mathbf{x})$ is *strictly convex* for $\mathbf{x} \in R^n$ if for any two points $\mathbf{x}, \mathbf{y} \in R^n$ and for all θ, $0 < \theta < 1$,

$$f\{\theta\mathbf{x}+(1-\theta\mathbf{y}\} < \theta f(\mathbf{x})+(1-\theta)f(\mathbf{y}). \tag{1.8}$$

It is *convex* if $<$ is replaced by $\leqslant$ in (1.8). This relation implies that the function has the behaviour sketched in Fig. 1.2.

Note that f need not be C^1 – for example a piecewise linear function of $\mathbf{x}$ can be convex; and that if the gradient $\mathbf{g}(\mathbf{x})$ exists, then

$$f(\mathbf{x}+\mathbf{h}) \geqslant f(\mathbf{x})+\mathbf{h}^{\mathrm{T}}\mathbf{g}(\mathbf{x}) \tag{1.9}$$

for all $\mathbf{h}$, if and only if f is convex. (Prove this.) It should be noted that a convex function can be defined when $\mathbf{x}$ is restricted to a set $X \subset R^n$ but that X must then be a convex set – see Section 1.2.

The connections between (4), (5) and (6) follow immediately from the definitions.

(a) If f is of class C^2 and is convex, then $\mathbf{G}(\mathbf{x})$ defined in (1.5) is positive semidefinite for all $\mathbf{x}$, and conversely.

(b) If $f(\mathbf{x})$ is quadratic, as in (1.7), and $\mathbf{G}$ is positive definite, then $f(\mathbf{x})$ is strictly convex, and conversely.

(c) If $f(\mathbf{x})$ is convex and has a strong local minimum, then this is unique and so it must be the global minimum.

The proof of (c) is given below – the student should prove (a) and (b) similarly.

Proof If $f(\mathbf{x})$ has a strong local minimum at $\mathbf{x}^*$ then from (1)

$$f(\mathbf{x}) > f(\mathbf{x}^*) \quad \text{for all } \mathbf{x} \text{ such that } \|\mathbf{x}-\mathbf{x}^*\| < \varepsilon.$$

For any other point $\boldsymbol{\xi}$, there exists a point $\bar{\mathbf{x}}$ on the line joining $\mathbf{x}^*, \boldsymbol{\xi}$ and in this ε-neighbourhood. That is, there exists $\bar{\mathbf{x}}$ such that

$$\bar{\mathbf{x}} = \bar{\theta}\boldsymbol{\xi} + (1-\bar{\theta})\mathbf{x}^* \quad \text{with} \quad 0 < \bar{\theta} < 1$$

and $\|\bar{\mathbf{x}} - \mathbf{x}^*\| < \varepsilon$, and so

$$f(\bar{\mathbf{x}}) > f(\mathbf{x}^*). \tag{1.10}$$

Now assume that $\mathbf{x}^*$ is not the global minimum. In this case there must exist a point $\boldsymbol{\xi}$ where $f(\boldsymbol{\xi}) < f(\mathbf{x}^*)$.

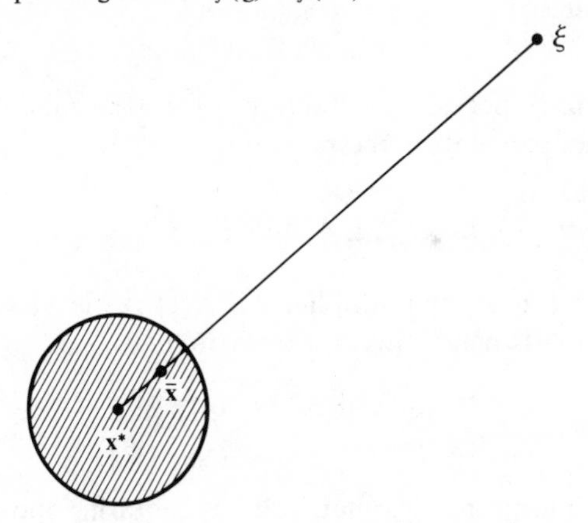

Fig. 1.3. Local and global minima.

Then since $f(\mathbf{x})$ is convex,

$$f(\bar{\mathbf{x}})=f\{\bar{\theta}\boldsymbol{\xi}+(1-\bar{\theta})\mathbf{x}^*\}\leqslant\bar{\theta}f(\boldsymbol{\xi})+(1-\bar{\theta})f(\mathbf{x}^*)<f(\mathbf{x}^*).$$

But this contradicts (1.10), and hence there is no such $\boldsymbol{\xi}$, i.e. $\mathbf{x}^*$ is the global minimum.

True convex functions occur rarely but they are important theoretically because functions are convex in the neighbourhood of a local minimum.

1.1.3 Equality constraints

A more general problem is to minimise $f(\mathbf{x})$ where the $\mathbf{x}$ satisfy m $(<n)$ equality constraints

$$c_i(\mathbf{x})=0, \quad i=1,2,\ldots,m \tag{1.11}$$

but are otherwise unrestricted. Admissible points are now confined to an $(n-m)$-dimensional subspace S determined by (1.11).

Necessary conditions If the $c_i\in C^1$, then $\mathbf{x}^*+\varepsilon\mathbf{h}\in S$, $\mathbf{x}^*\in S$ implies

$$\mathbf{h}^{\mathrm{T}}\mathbf{a}_i(\mathbf{x}^*)=0 \tag{1.12}$$

where $\mathbf{a}_i(\mathbf{x})$ is the gradient vector $\nabla c_i(\mathbf{x})$. If the $\mathbf{a}_i(\mathbf{x}^*)$ are independent, (1.12) defines the tangent plane to S at $\mathbf{x}^*$. If $\mathbf{x}^*$ is a local minimum point for $f(\mathbf{x})$, subject to $\mathbf{x}\in S$, then for all $\mathbf{h}$ satisfying (1.12)

$$f(\mathbf{x}^*+\varepsilon\mathbf{h})\geqslant f(\mathbf{x}^*)$$

and this implies

$$\mathbf{h}^{\mathrm{T}}\mathbf{g}(\mathbf{x}^*)=0 \tag{1.13}$$

for all such $\mathbf{h}$. A necessary condition for a local minimum, if the $\mathbf{a}_i(\mathbf{x}^*)$ are independent, is then

$$\mathbf{g}(\mathbf{x}^*)=\sum_{i=1}^{m}\lambda_i^*\mathbf{a}_i(\mathbf{x}^*), \quad \mathbf{x}^*\in S \tag{1.14}$$

where the λ_i^* are scalar multipliers; if $\mathbf{A}(\mathbf{x})$ is the $m\times n$ matrix whose i-th row is $\mathbf{a}_i^{\mathrm{T}}(\mathbf{x})$, this can be written

$$\left.\begin{aligned}\mathbf{g}(\mathbf{x}^*)&=\{\mathbf{A}(\mathbf{x}^*)\}^{\mathrm{T}}\boldsymbol{\lambda}^*, &&\text{where } \boldsymbol{\lambda}^*\in R^m\\ \mathbf{c}(\mathbf{x}^*)&=0 &&\text{where } \mathbf{c}\in R^m.\end{aligned}\right. \tag{1.15}$$

This can be interpreted geometrically as requiring the gradient vector to be normal to the tangent plane at $\mathbf{x}^*$ to S.

These conditions need not hold if the $\mathbf{a}_i(\mathbf{x}^*)$ are not independent, that is if the rank of $\mathbf{A}(\mathbf{x}^*) < m$ – see Problem 8. This is the abnormal situation which qualifies the theoretical statements in both equality- and inequality-constrained problems, but which can generally be avoided in practice, here by perturbing the constraints. We shall note when a qualification of this sort is required but will normally assume that the general conditions are adequate.

Implicit function approach An alternative derivation of the conditions (1.15) is by using the Implicit Function Theorem. This also gives information directly about the meaning of the λ_i^*.

Suppose $c_i(\mathbf{x}^*) = 0$, and the $m \times m$ minor of $\mathbf{A}^T(\mathbf{x})$,

$$\mathbf{C}^{(1)}(\mathbf{x}) = \begin{bmatrix} \dfrac{\partial c_1}{\partial x_1} & \cdots & \dfrac{\partial c_m}{\partial x_1} \\ \vdots & & \\ \dfrac{\partial c_1}{\partial x_m} & \cdots & \dfrac{\partial c_m}{\partial x_m} \end{bmatrix} \tag{1.16}$$

is nonsingular at $\mathbf{x}^*$. (Since the $\mathbf{a}_i(\mathbf{x}^*)$ are assumed linearly independent, such a $\mathbf{C}^{(1)}$ exists for some selection of m x's from the n, so suppose these m have been if necessary renumbered to become $x_1, x_2, \ldots x_m$.) Then the Implicit Function Theorem states that there exist unique functions X_i, $i = 1, 2, \ldots, m$, of the remaining $(n-m)$ x's such that $X_i \in C^1$ and the constraints are identically satisfied in an open set containing $\mathbf{x}^*$, that is

$$c_i\{X_1(x_{m+1}, \ldots, x_n), X_2(x_{m+1}, \ldots, x_n), \ldots, X_m(x_{m+1}, \ldots, x_n), \\ x_{m+1}, \ldots, x_n\} = 0 \tag{1.17}$$

for all $x_{m+1}, \ldots, x_n$ in this set; and also the point $\mathbf{x}^*$ is the minimum point of the function f regarded as a function of the $n-m$ unrestricted variables $x_{m+1}, \ldots, x_n$ in this open set. (See any Analysis text for a proof of this.)

It follows from (1.17) that at $\mathbf{x}^*$

$$\frac{\partial c_i}{\partial x_1}\frac{\partial X_1}{\partial x_j} + \frac{\partial c_i}{\partial x_2}\frac{\partial X_2}{\partial x_j} + \cdots + \frac{\partial c_i}{\partial x_m}\frac{\partial X_m}{\partial x_j} + \frac{\partial c_i}{\partial x_j} = 0, \quad 1 \leqslant i \leqslant m, \; m+1 \leqslant j \leqslant n, \tag{1.18}$$

and since

$$f(\mathbf{x}) = f\{X_1(x_{m+1}, \ldots, x_n), X_2(x_{m+1}, \ldots, x_n), \ldots,$$
$$X_m(x_{m+1}, \ldots, x_n), x_{m+1}, \ldots, x_n\}, \quad (1.19)$$

then from (1.3) also at $\mathbf{x}^*$,

$$\frac{\partial f}{\partial x_1}\frac{\partial X_1}{\partial x_j} + \frac{\partial f}{\partial x_2}\frac{\partial X_2}{\partial x_j} + \cdots + \frac{\partial f}{\partial x_m}\frac{\partial X_m}{\partial x_j} + \frac{\partial f}{\partial x_j} = 0, \quad m+1 \leqslant j \leqslant n. \quad (1.20)$$

Hence defining the $(n-m) \times m$ matrices $\mathbf{V}$, $\mathbf{C}^{(2)}$ by

$$\mathbf{V} = \begin{bmatrix} \dfrac{\partial X_1}{x_{m+1}} & \cdots & \dfrac{\partial X_m}{x_{m+1}} \\ \vdots & & \\ \dfrac{\partial X_1}{\partial x_n} & \cdots & \dfrac{\partial X_m}{\partial x_n} \end{bmatrix} \quad \mathbf{C}^{(2)} = \begin{bmatrix} \dfrac{\partial c_1}{\partial x_{m+1}} & \cdots & \dfrac{\partial c_m}{\partial x_{m+1}} \\ \vdots & & \\ \dfrac{\partial c_1}{\partial x_n} & \cdots & \dfrac{\partial c_m}{\partial x_n} \end{bmatrix}, \quad (1.21)$$

all derivatives being evaluated at $\mathbf{x}^*$, and also partitioning the vector $\mathbf{g}(\mathbf{x}^*)$ into

$$\mathbf{g}^{(1)} = \begin{bmatrix} g_1(\mathbf{x}^*) \\ \vdots \\ g_m(\mathbf{x}^*) \end{bmatrix} \quad \text{and} \quad \mathbf{g}^{(2)} = \begin{bmatrix} g_{m+1}(\mathbf{x}^*) \\ \vdots \\ g_n(\mathbf{x}^*) \end{bmatrix}$$

conditions (1.18), (1.20) become

$$\mathbf{V}\mathbf{C}^{(1)} + \mathbf{C}^{(2)} = 0,$$
$$\mathbf{V}\mathbf{g}^{(1)} + \mathbf{g}^{(2)} = 0,$$

and since $\mathbf{C}^{(1)}$ is nonsingular, $\mathbf{V} = -\mathbf{C}^{(2)}\{\mathbf{C}^{(1)}\}^{-1}$ and

$$-\mathbf{C}^{(2)}\{\mathbf{C}^{(1)}\}^{-1}\mathbf{g}^{(1)} + \mathbf{g}^{(2)} = 0. \quad (1.22)$$

Now write

$$\{\mathbf{C}^{(1)}\}^{-1}\mathbf{g}^{(1)} = \boldsymbol{\lambda}^* \quad (1.23)$$

where $\boldsymbol{\lambda}^{*\mathrm{T}}$ is the m-vector with components $(\lambda_1^*, \lambda_2^*, \ldots, \lambda_m^*)$, and (1.22), (1.23) become

$$\begin{bmatrix} \mathbf{C}^{(1)} \\ \mathbf{C}^{(2)} \end{bmatrix} \boldsymbol{\lambda}^* = \begin{bmatrix} \mathbf{g}^{(1)} \\ \mathbf{g}^{(2)} \end{bmatrix},$$

or, as in (1.15), $\mathbf{g}(\mathbf{x}^*) = \{\mathbf{A}(\mathbf{x}^*)\}^{\mathrm{T}}\boldsymbol{\lambda}^*$.

Sensitivity to constraint position Now take the constraints in the form

$$c_i(\mathbf{x}) = b_i. \tag{1.24}$$

The same analysis gives function X_i now dependent on $x_{m+1}, \ldots, x_n$ and $b_1, \ldots, b_m$, so that

$$c_i\{X_1, \ldots, X_m, x_{m+1}, \ldots, x_n\} = b_i$$

and as before,

$$f\{X_1, \ldots, X_m, x_{m+1}, \ldots, x_n\}$$

has a minimum at $\mathbf{x}^*$ with respect to $x_{m+1}, \ldots, x_n$. However, now $\mathbf{x}^*$ is dependent on $\mathbf{b}$, and so is $f(\mathbf{x}^*)$, the minimum value. Since the X_i now are differentiable functions of the b's as well as the x's, we may write similar equations to (1.18), (1.20): at $\mathbf{x}^*$,

$$\begin{aligned}\frac{\partial c_i}{\partial x_1}\frac{\partial X_1}{\partial b_j}+\frac{\partial c_i}{\partial x_2}\frac{\partial X_2}{\partial b_j}+\cdots+\frac{\partial c_i}{\partial x_m}\frac{\partial X_m}{\partial b_j}&\\ +\frac{\partial c_i}{\partial x_{m+1}}\frac{\partial x^*_{m+1}}{\partial b_j}+\cdots+\frac{\partial c_i}{\partial x_n}\frac{\partial x^*_n}{\partial b_j}&=\delta_{ij},\\ 1\leqslant i\leqslant m,\quad 1\leqslant j\leqslant m&\end{aligned} \tag{1.25}$$

and

$$\begin{aligned}\frac{\partial f^*}{\partial b_j}=\frac{\partial f^*}{\partial x_1}\frac{\partial X_1}{\partial b_j}+\cdots+\frac{\partial f^*}{\partial x_m}\frac{\partial X_m}{\partial b_j}&\\ +\frac{\partial f^*}{\partial x_{m+1}}\frac{\partial x^*_{m+1}}{\partial b_j}+\cdots+\frac{\partial f^*}{\partial x_n}\frac{\partial x^*_n}{\partial b_j},&\quad 1\leqslant j\leqslant m.\end{aligned} \tag{1.26}$$

Defining the $m\times m$ and $m\times(n-m)$ matrices $\mathbf{B}^{(1)}$ and $\mathbf{B}^{(2)}$ respectively as

$$\mathbf{B}^{(1)}=\begin{bmatrix}\dfrac{\partial X_1}{\partial b_1} & \cdots & \dfrac{\partial X_m}{\partial b_1}\\ \vdots & & \\ \dfrac{\partial X_1}{\partial b_m} & \cdots & \dfrac{\partial X_m}{\partial b_m}\end{bmatrix},\quad \mathbf{B}^{(2)}=\begin{bmatrix}\dfrac{\partial x^*_{m+1}}{\partial b_1} & \cdots & \dfrac{\partial x^*_n}{\partial b_1}\\ \vdots & & \\ \dfrac{\partial x^*_{m+1}}{\partial b_m} & \cdots & \dfrac{\partial x^*_n}{\partial b_m}\end{bmatrix}, \tag{1.27}$$

all derivatives being evaluated at $\mathbf{x}^*$, then (1.25), (1.26) together with (1.16), (1.21) give

$$\mathbf{B}^{(1)}\mathbf{C}^{(1)}+\mathbf{B}^{(2)}\mathbf{C}^{(2)}=\mathbf{I}$$

and

$$\frac{\partial f^*}{\partial \mathbf{b}} = \mathbf{B}^{(1)}\mathbf{g}^{(1)} + \mathbf{B}^{(2)}\mathbf{g}^{(2)} \tag{1.28}$$

where $(\partial f^*/\partial \mathbf{b})^{\mathrm{T}}$ is the m-vector

$$\left(\frac{\partial f^*}{\partial b_1}, \frac{\partial f^*}{\partial b_2}, \ldots, \frac{\partial f^*}{\partial b_m}\right).$$

Substituting in (1.28) from (1.22), (1.23) gives

$$\frac{\partial f^*}{\partial \mathbf{b}} = \boldsymbol{\lambda}^*. \tag{1.29}$$

Thus λ_i^* is a multiplier which shows the sensitivity of the value f^* to the position of the i-th constraint.

Lagrange multiplier formulation An alternative approach to equality-constrained minimisation is by Lagrange multipliers. The function $f(\mathbf{x})$ and the constraint functions $c_i(\mathbf{x})$ are combined using multipliers λ_i to form the Lagrangian

$$L(\mathbf{x}, \boldsymbol{\lambda}) = f(\mathbf{x}) - \sum_{i=1}^{m} \lambda_i c_i(\mathbf{x}) = f(\mathbf{x}) - \{\mathbf{c}(\mathbf{x})\}^{\mathrm{T}}\boldsymbol{\lambda}. \tag{1.30}$$

Then the constrained problem:

$$\text{minimise } f(\mathbf{x}) \text{ over } \mathbf{x} \in S$$

is replaced by the unconstrained problem:

find the critical value of L for $\mathbf{x}$, $\boldsymbol{\lambda}$ unconstrained.

This follows because the conditions for L to have a critical value at $\mathbf{x}^*$, $\boldsymbol{\lambda}^*$ are

$$\nabla_{\mathbf{x}} L = 0, \quad \nabla_{\boldsymbol{\lambda}} L = 0 \tag{1.31}$$

or

$$\mathbf{g}(\mathbf{x}^*) - \{\mathbf{A}(\mathbf{x}^*)\}^{\mathrm{T}}\boldsymbol{\lambda}^* = 0, \quad \mathbf{c}(\mathbf{x}^*) = 0$$

which are the same conditions as (1.15). Hence the $\boldsymbol{\lambda}^*$ which have already appeared, in (1.15) and, explicitly, in (1.23), can be identified as the optimum Lagrange multipliers. Note that the critical value of the Lagrangian, $L(\mathbf{x}^*, \boldsymbol{\lambda}^*)$, is just $f(\mathbf{x}^*)$ since $\mathbf{c}(\mathbf{x}^*)$ is zero.

Sufficiency conditions The critical point $\mathbf{x}^*$, satisfying (1.15), is a minimum if and only if

$$f(\mathbf{x}^*+\mathbf{h}) \geqslant f(\mathbf{x}^*) \quad \text{for any } \mathbf{h} \text{ such that } \mathbf{x}^*+\mathbf{h} \in S.$$

Alternatively, f as defined in (1.19) has a minimum $\mathbf{x}^*$ if (1.14) holds, and if also the Hessian of f with respect to the $(n-m)$ variables $x_{m+1}, \ldots, x_n$ is strictly positive definite at $\mathbf{x}^*$.

A simple though more restrictive condition is that the Lagrangian $L(\mathbf{x}, \boldsymbol{\lambda}^*)$ should be a convex function of $\mathbf{x}$; or, equivalently, that if $f, c_i \in C^2$, then the Hessian $\mathbf{F}(\mathbf{x}^*, \boldsymbol{\lambda}^*)$ should be positive definite, where

$$\mathbf{F}(\mathbf{x}, \boldsymbol{\lambda}) = \mathbf{G}(\mathbf{x}) - \sum_{i=1}^{m} \lambda_i \mathbf{C}_i(\mathbf{x}), \tag{1.32}$$

$\mathbf{C}_i(\mathbf{x})$ being the Hessian of $c_i(\mathbf{x})$. This is sufficient to ensure that a critical point, satisfying (1.15), is a minimum point for $L(\mathbf{x}, \boldsymbol{\lambda}^*)$ in $\mathbf{x}$, and is also a minimum point for $f(\mathbf{x})$ subject to $\mathbf{x} \in S$.

Proof

$$L(\mathbf{x}^*+\mathbf{h}, \boldsymbol{\lambda}^*) = L(\mathbf{x}^*, \boldsymbol{\lambda}^*) + \mathbf{h}^{\mathrm{T}} \nabla_x L(\mathbf{x}^*, \boldsymbol{\lambda}^*) + \tfrac{1}{2} \mathbf{h}^{\mathrm{T}} \mathbf{F}(\mathbf{x}^*, \boldsymbol{\lambda}^*) \mathbf{h} + O(\|\mathbf{h}\|^3)$$

and so if (1.15) holds and $\mathbf{F}$ is positive definite,

$$L(\mathbf{x}, \boldsymbol{\lambda}^*) \geqslant L(\mathbf{x}^*, \boldsymbol{\lambda}^*), \quad \text{all } \mathbf{x};$$

hence

$$L(\mathbf{x}, \boldsymbol{\lambda}^*) \geqslant L(\mathbf{x}^*, \boldsymbol{\lambda}^*), \quad \text{for } \mathbf{x} \in S$$

so

$$f(\mathbf{x}) - \{\mathbf{c}(\mathbf{x})\}^{\mathrm{T}} \boldsymbol{\lambda}^* \geqslant f(\mathbf{x}^*) - \{\mathbf{c}(\mathbf{x}^*)\}^{\mathrm{T}} \boldsymbol{\lambda}^* \quad \text{for } \mathbf{x} \in S,$$

and since $\mathbf{x}^* \in S$, this reduces to

$$f(\mathbf{x}) \geqslant f(\mathbf{x}^*) \quad \text{for } \mathbf{x} \in S$$

as required.

General conditions, corresponding to (1.6), are (1.15) and

$$\mathbf{h}^{\mathrm{T}} \mathbf{F}(\mathbf{x}^*, \boldsymbol{\lambda}^*) \mathbf{h} > 0$$

for h such that $\mathbf{h}^{\mathrm{T}} \mathbf{a}_i(\mathbf{x}^*) = 0$. It is now $\mathbf{F}(\mathbf{x}^*, \boldsymbol{\lambda}^*)$ which is required to be positive definite, but only on a subspace, the tangent plane to S at $\mathbf{x}^*$.

Note that the conditions for a local minimum are the same for $\mathbf{x}$ restricted to a neighbourhood of $\mathbf{x}^*$; that is, if the Lagrangian is locally convex, then its critical point is a local minimum and is also a local minimum of $f(\mathbf{x})$.

1.2 Inequality constraints

1.2.1 General considerations

If $f(\mathbf{x})$ is to be minimised subject to restrictions on $\mathbf{x}$ of the form

$$c_i(\mathbf{x}) \geqslant 0, \quad i = 1, 2, \ldots, m, \tag{1.33}$$

then this corresponds to the problem

$$\text{minimise } f(\mathbf{x}) \text{ subject to } \mathbf{x} \in X,$$

where X is defined by (1.33) and is the *solution set* or *feasible set* for the problem.

The definitions of a local and global minimum still apply as in Section 1.1.1 and the necessary conditions for a local minimum may be written generally, including the two cases already described as special cases, in the following form:

$\mathbf{x}^*$ is a local minimum of $f(\mathbf{x})$ subject to $\mathbf{x} \in X$ if
(a) $\mathbf{x}^* \in X$ and
(b) there is no vector $\mathbf{p}$ such that
 (i) $\mathbf{x} = \mathbf{x}^* + \alpha\mathbf{p} \in X$ for all α in the range $0 \leqslant \alpha \leqslant \bar{\alpha}$, $\bar{\alpha} > 0$, and

 (ii) $\mathbf{g}(\mathbf{x})^{\mathrm{T}}\mathbf{p} < 0$ for all such $\mathbf{x}$. (1.34)

The geometrical significance of this condition is clear; any move from $\mathbf{x}^*$ to an adjacent feasible point $\mathbf{x}^* + \alpha\mathbf{p}$ must not be downhill, i.e. must not have a negative component $\mathbf{g}(\mathbf{x})^{\mathrm{T}}\mathbf{p}$. There are two cases to be considered. First, if $\mathbf{x}^*$ is not on the boundary of X, that is if

$$c_i(\mathbf{x}^*) > 0, \quad i = 1, 2, \ldots, m,$$

then there is a neighbourhood of $\mathbf{x}^*$ within X and hence by the same argument as before it is necessary that

$$\mathbf{g}(\mathbf{x}^*) = 0. \tag{1.35}$$

Secondly, suppose $\mathbf{x}^*$ is on the boundary, that is

$$c_i(\mathbf{x}^*) = 0 \quad \text{for some } i, i \in I. \tag{1.36}$$

The set I is called the active set and is in general unknown, except in the limiting case (Section 1.1.3) where all the constraints are equalities and so are all necessarily active.

It would be theoretically possible to take all possible subsets I_ν containing $0, 1, 2, \ldots, m$ of the m constraints as active sets and to write down necessary conditions for the corresponding $\mathbf{x}_\nu^*$ as in Section 1.1.3. Many such $\mathbf{x}_\nu^*$ would turn out to be infeasible, but any feasible $\mathbf{x}_\nu^*$ would be local stationary points for the problem (1.33). This is not generally a useful form of condition, since there are a very large number of such subsets I_ν when m is large; however, many methods for finding local minima involve some informed selection of active sets. In particular if $f(\mathbf{x})$, $c_i(\mathbf{x})$ are linear in $\mathbf{x}$, the Simplex method of solution operates this way but with the very important simplification that the number of active constraints is known – see Chapter 3.

1.2.2 Necessary conditions, Kuhn–Tucker form

The conditions can be obtained by similar arguments to those in Section 1.1: we develop first those appropriate to a normal point, $\mathbf{x}$, defining this as a point at which the gradients $\mathbf{a}_i(\mathbf{x})$ of the active constraints $i \in I$ are linearly independent. A note to cover the case of an abnormal point is added in Section 1.2.4.

If

$$\begin{aligned} c_i(\mathbf{x}^*) &= 0, \quad i \in I \\ c_i(\mathbf{x}^*) &> 0 \quad \text{otherwise,} \end{aligned} \tag{1.37}$$

then as in (1.14), if the $\mathbf{a}_i(\mathbf{x}^*)$, $i \in I$, are linearly independent, and $\mathbf{x}^*$ minimises $f(\mathbf{x})$ subject to (1.37),

$$\mathbf{g}(\mathbf{x}^*) = \sum_{i \in I} \lambda_i^* \mathbf{a}_i(\mathbf{x}^*)$$

and, if we define $\lambda_i^* = 0$ for $i \notin I$, this can be written, as in (1.15),

(i) $\mathbf{g}(\mathbf{x}^*) = \{\mathbf{A}(\mathbf{x}^*)\}^{\mathrm{T}} \boldsymbol{\lambda}^*$

together with

(ii) $\lambda_i^* c_i(\mathbf{x}^*) = 0$ (1.38)

and

(iii) $\mathbf{c}(\mathbf{x}^*) \geqslant 0$.

In addition, a necessary condition at a minimum is

(iv) $\boldsymbol{\lambda}^* \geqslant 0$ (1.39)

For if $\mathbf{x}^*$ is a minimum, then $f(\mathbf{x}^*+\mathbf{h}) \geqslant f(\mathbf{x}^*)$ for any feasible $\mathbf{h}$. Condition (i) above is necessary since, as before, this must hold for all $\mathbf{h}$ such that $\mathbf{h}^T\mathbf{a}_i(\mathbf{x}^*)=0$ for $i \in I$, that is for displacements consistent with the active constraints at $\mathbf{x}^*$. However, we now need also to consider moves away from the constraint boundaries into the feasible region. Let $\mathbf{h}$ be such a move, for which all constraints except the k-th remain active; then

$$\mathbf{h}^T\mathbf{a}_i(\mathbf{x}^*)=0 \quad \text{for } i \in I, i \neq k,$$

and

$$c_k(\mathbf{x}^*+\mathbf{h}) \geqslant c_k(\mathbf{x}^*)=0$$

so that

$$\mathbf{h}^T\mathbf{a}_k(\mathbf{x}^*) \geqslant 0.$$

For such $\mathbf{h}$

$$f(\mathbf{x}^*+\mathbf{h})=f(\mathbf{x}^*)+\mathbf{h}^T\mathbf{g}(\mathbf{x}^*)$$

to first order in $\|\mathbf{h}\|$, and substituting from (1.14),

$$\begin{aligned} f(\mathbf{x}^*+\mathbf{h}) &= f(\mathbf{x}^*)+\mathbf{h}^T\sum_{i\in I}\lambda_i^*\mathbf{a}_i(\mathbf{x}^*) \\ &= f(\mathbf{x}^*)+\mathbf{h}^T\lambda_k^*\mathbf{a}_k(\mathbf{x}^*) \end{aligned}$$

from the condition on $\mathbf{h}$, and so it is necessary that $\lambda_k^* \geqslant 0$.

Conditions (i)–(iv) are the Kuhn–Tucker conditions which hold at a minimum point. We can prove that they are also sufficient for there to be no vector $\mathbf{p}$ satisfying simultaneously

$$\mathbf{p}^T\mathbf{g}(\mathbf{x}^*)<0, \quad \mathbf{p}^T\mathbf{a}_i(\mathbf{x}^*) \geqslant 0,$$

that is they are first-order sufficient conditions for $\mathbf{x}^*$ to be a minimum. This follows from Farkas' Lemma which is given in the next section. Second-order sufficient conditions are discussed in Section 1.2.5.

It may be noted that this condition (1.39) on $\boldsymbol{\lambda}^*$ follows also from (1.29). An increase in any b, say b_k, moves the corresponding boundary $c_k(\mathbf{x})=b_k$ into the interior of the feasible set (Fig. 1.4). If the k-th is an active constraint, the optimal value $f(\mathbf{x}^*)$ may change, and since the feasible region has been reduced, such change must be non-negative, while a change in an inactive constraint will not alter the optimum value. Hence

$$\frac{\partial f^*}{\partial b_k}=\lambda_k^*, \quad \lambda_k^* \geqslant 0 \text{ for } i \in I, \quad \lambda_k^*=0 \text{ for } i \notin I.$$

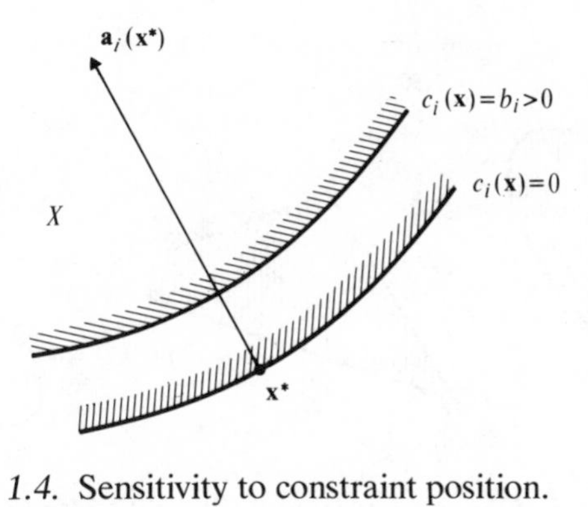

Fig. 1.4. Sensitivity to constraint position.

A common restriction on $\mathbf{x}$ in a constrained problem is to require some x_j to be non-negative; this constraint may be used directly in the form given above, but it is sometimes convenient to state it implicitly. Thus in minimising $f(\mathbf{x})$ subject to $c_i(\mathbf{x}) \geqslant 0$, $i=1,\ldots,m$ and $x_j \geqslant 0$, $j \in J$, the Kuhn–Tucker conditions are, with multipliers $\lambda_i^*, \mu_j^* \geqslant 0$

$$g_j(\mathbf{x}^*) - \sum_{i=1}^{m} \lambda_i^* a_{ij}(\mathbf{x}^*) - \mu_j^* = 0, \quad j \in J$$

$$g_j(\mathbf{x}^*) - \sum_{i=1}^{m} \lambda_i^* a_{ij}(\mathbf{x}^*) = 0, \quad j \notin J$$

which may be combined, eliminating the μ_j^*, to give

$$\mathbf{g}(\mathbf{x}^*) - \sum_{i=1}^{m} \lambda_i^* \mathbf{a}_i(\mathbf{x}^*) \geqslant 0, \quad \mathbf{x}^{*\mathrm{T}}\left\{\mathbf{g}(\mathbf{x}^*) - \sum_{i=1}^{m} \lambda_i^* \mathbf{a}_i(\mathbf{x}^*)\right\} = 0. \tag{1.40}$$

1.2.3 Convexity, Farkas' Lemma

Before developing the argument we give some definitions.

(a) Convex set A set K is convex if for any two points $\mathbf{x}_1$, $\mathbf{x}_2$ in K, all points on the line segment joining $\mathbf{x}_1$ and $\mathbf{x}_2$ are also in K: that is $\mathbf{x}_1$, $\mathbf{x}_2$ in K implies $\mathbf{x} = \theta\mathbf{x}_1 + (1-\theta)\mathbf{x}_2$, $0<\theta<1$, is also in K. Figure 1.5 shows convex sets (i) and (ii) and a nonconvex set (iii).

(b) Convex cone A set K is a convex cone if it is a convex set and if $\mathbf{x} \in K$ implies $\alpha\mathbf{x} \in K$ for α a non-negative scalar.

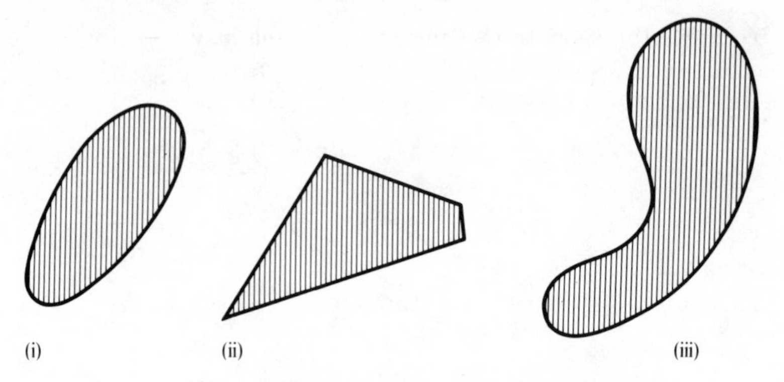

Fig. 1.5. Convex sets (i) and (ii), and a nonconvex set (iii).

Farkas' Lemma states that:
given the vectors $\mathbf{a}_i$, $i = 1, 2, \ldots, t$, and a vector $\mathbf{g}$, there is no vector $\mathbf{p}$ satisfying simultaneously

$$\mathbf{g}^T\mathbf{p} < 0, \quad \mathbf{a}_i^T\mathbf{p} \geqslant 0, \quad i = 1, 2, \ldots, t \tag{1.41}$$

if and only if $\mathbf{g}$ is expressible in the form

$$\mathbf{g} = \sum_{i=1}^{t} \lambda_i \mathbf{a}_i \quad \text{with } \lambda_i \geqslant 0. \tag{1.42}$$

The set I here is the set $i = 1, 2, \ldots, t$.
Proof The set

$$K = \left\{\mathbf{v} \mid \mathbf{v} = \sum_{i=1}^{t} \lambda_i \mathbf{a}_i, \lambda_i \geqslant 0\right\}$$

is a convex cone: definitions (a) and (b) can be verified directly.

If $\mathbf{g} \in K$, then $\mathbf{g}^T\mathbf{p} = \sum_{i=1}^t \lambda_i \mathbf{a}_i^T\mathbf{p} \geqslant 0$ by (1.41), (1.42). Hence the first part of the lemma is proved.

Conversely, suppose $\mathbf{g} \notin K$, i.e. suppose $\mathbf{g}$ cannot be expressed as $\sum_{i=1}^t \lambda_i \mathbf{a}_i$. There are two cases to consider. First, suppose that $\mathbf{g}^T\mathbf{a}_i \leqslant 0$ for all i; then clearly $\mathbf{p} = -\mathbf{g}$ satisfies $\mathbf{g}^T\mathbf{p} < 0$, $\mathbf{p}^T\mathbf{a}_i \geqslant 0$. Secondly, suppose that $\mathbf{g}^T\mathbf{a}_i > 0$ for some or all i. Then if $\mathbf{h}$ is the vector $\in K$ for which the norm $\|\mathbf{h} - \mathbf{g}\|$ is a minimum, we can show that

(a) $\mathbf{p} = \mathbf{h} - \mathbf{g}$ satisfies $\mathbf{p}^T\mathbf{v} \geqslant 0$ for all $\mathbf{v} \in K$,
and so $\mathbf{p}^T\mathbf{a}_i \geqslant 0$, all i; (1.43)

(b) $\mathbf{p}^T\mathbf{h} = 0$, so that $\mathbf{p}^T\mathbf{g} = -\|\mathbf{p}\|^2 < 0$. (1.44)

To prove (a), suppose that this is not so, that $\mathbf{p}^{\mathrm{T}}\mathbf{v} = -d$ for some $d > 0$, and some $\mathbf{v} \in K$; then since K is a convex cone, $\mathbf{h} + d\mathbf{v} \in K$, and we may take $\|\mathbf{v}\| = 1$. But

$$\begin{aligned}\|\mathbf{h} + d\mathbf{v} - \mathbf{g}\|^2 &= \|\mathbf{p}\|^2 + 2d\mathbf{p}^{\mathrm{T}}\mathbf{v} + d^2 \\ &= \|\mathbf{p}\|^2 - d^2 < \|\mathbf{p}\|^2,\end{aligned}$$

which is a contradiction, since $\mathbf{h}$ is the vector in K with minimum norm $\|\mathbf{h} - \mathbf{g}\|$.

To prove (b), set $\mathbf{h} = a\hat{\mathbf{h}}$ and consider $\|a\hat{\mathbf{h}} - \mathbf{g}\|$, where $\|\hat{\mathbf{h}}\| = 1$; since

$$\|a\hat{\mathbf{h}} - \mathbf{g}\|^2 = a^2 - 2a\hat{\mathbf{h}}^{\mathrm{T}}\mathbf{g} + \|\mathbf{g}\|^2$$

this is smallest, for fixed $\hat{\mathbf{h}}$, when $a = \hat{\mathbf{h}}^{\mathrm{T}}\mathbf{g}$, and then has the value $\|\mathbf{g}\|^2 - (\hat{\mathbf{h}}^{\mathrm{T}}\mathbf{g})^2$. Hence if $\hat{\mathbf{h}}$ is the unit vector $\in K$ which maximises $\hat{\mathbf{h}}^{\mathrm{T}}\mathbf{g}$, then $\mathbf{h} = (\hat{\mathbf{h}}^{\mathrm{T}}\mathbf{g})\hat{\mathbf{h}}$ minimises $\|\mathbf{h} - \mathbf{g}\|$, and so

$$\mathbf{p}^{\mathrm{T}}\mathbf{h} = (\mathbf{h} - \mathbf{g})^{\mathrm{T}}\mathbf{h} = (\mathbf{h}^{\mathrm{T}}\mathbf{g})^2 - (\mathbf{h}^{\mathrm{T}}\mathbf{g})^2 = 0.$$

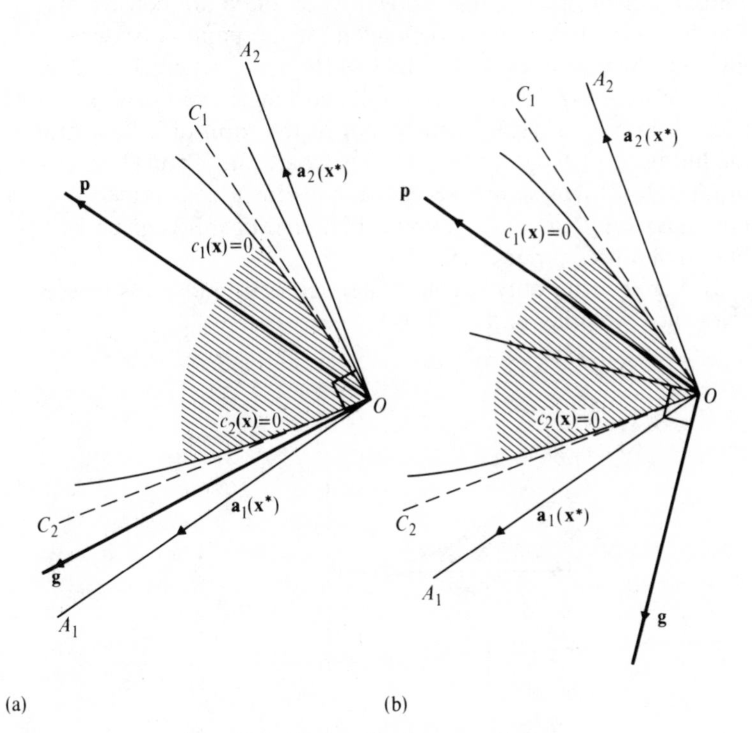

Fig. 1.6. Kuhn–Tucker conditions.

Hence if $\mathbf{g}$ is not expressible as $\sum \lambda_i \mathbf{a}_i$, $\lambda_i \geqslant 0$, there exists $\mathbf{p}$ such that

$$\mathbf{g}^{\mathrm{T}}\mathbf{p} < 0, \quad \mathbf{a}_i^{\mathrm{T}}\mathbf{p} \geqslant 0.$$

The geometrical significance of this can be sketched in two dimensions (Fig. 1.6). In case (a), $\mathbf{g}(\mathbf{x}^*)$ is a non-negative combination of $\mathbf{a}_1(\mathbf{x}^*)$ and $\mathbf{a}_2(\mathbf{x}^*)$, that is it lies within the cone A_1OA_2. It can be seen that a move from $\mathbf{x}^*$ to any adjacent feasible point will lie within the cone C_1OC_2 where C_1O, C_2O are the linearised versions of $c_1(\mathbf{x}) = 0$, $c_2(\mathbf{x}) = 0$ near $\mathbf{x}^*$, and that any such move has a positive component along $\mathbf{g}(\mathbf{x}^*)$. In case (b), $\mathbf{g}(\mathbf{x}^*)$ lies outside the cone A_1OA_2, and a move from $\mathbf{x}^*$ to an adjacent point along $\mathbf{p}$ is such that $\mathbf{g}^{\mathrm{T}}\mathbf{p} < 0$.

1.2.4 Abnormal point

The argument above breaks down when there are constraints whose gradients are not independent at the point considered. We give an illustration only of this case (Fig. 1.7) where it is clear that the necessary conditions (1.42) no longer hold. The general specification as in (1.33) can be put in the form of a "constraint qualification" which covers all cases where the Kuhn–Tucker conditions (1.38), (1.39) are necessary ones for a constrained minimum. See Note 1.2. A good treatment of this is given by Fiacco and McCormick (1968).

In Fig. 1.7 the only feasible direction is $\mathbf{p}$, and $\mathbf{g}^{\mathrm{T}}\mathbf{p} < 0$, $\mathbf{a}_1^{\mathrm{T}}\mathbf{p} = 0$, $\mathbf{a}_2^{\mathrm{T}}\mathbf{p} = 0$ so violating (1.41), (1.42).

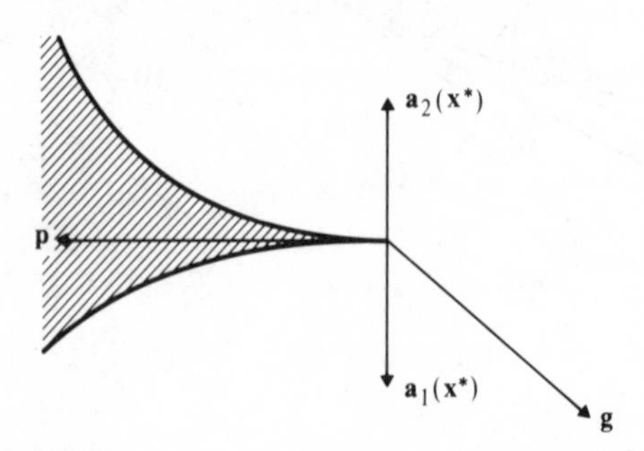

Fig. 1.7. Minimum where Kuhn–Tucker conditions do not hold.

1.2.5 Sufficient conditions

Convexity considerations may be used to establish some general results and in particular to establish conditions about global minima. Recall the definitions of a convex function, and add that of a concave function:

$f(\mathbf{x})$ is convex (concave) over a convex set X if for any two points $\mathbf{x}_1$, $\mathbf{x}_2$ in X and for all θ, $0 \leqslant \theta \leqslant 1$

$$f\{\theta\mathbf{x}_1+(1-\theta)\mathbf{x}_2\} \leqslant \theta f(\mathbf{x}_1)+(1-\theta)f(\mathbf{x}_2).$$
$$(\geqslant)$$

Note that both properties have to be considered over a convex set X since we clearly require that $\theta\mathbf{x}_1+(1-\theta)\mathbf{x}_2$ is in the set if $\mathbf{x}_1$ and $\mathbf{x}_2$ are. Note also that if $f(\mathbf{x})$ is convex, then $-f(\mathbf{x})$ is concave, and that if $f(\mathbf{x})$ is linear, it is concave and convex over the whole of R^n. Now we can establish some important properties.

(1) If $c_i(\mathbf{x})$ is a concave function over R^n, then the set of points X satisfying $c_i(\mathbf{x}) \geqslant 0$, $i=1, 2, \ldots, m$, is a convex set.

Proof Let $\mathbf{x}_1$, $\mathbf{x}_2$ be points such that $c_i(\mathbf{x}_1) \geqslant 0$, $c_i(\mathbf{x}_2) \geqslant 0$, $i=1, 2, \ldots, m$. Then for any i, $i=1, 2, \ldots, m$,

$$\begin{aligned} c_i\{\theta\mathbf{x}_1+(1-\theta)\mathbf{x}_2\} &\geqslant \theta c_i(\mathbf{x}_1)+(1-\theta)c_i(\mathbf{x}_2) \quad \text{(by concavity)} \\ &\geqslant 0 \quad \text{(since } \theta, 1-\theta \geqslant 0) \end{aligned}$$

and so $\theta\mathbf{x}_1+(1-\theta)\mathbf{x}_2$ is in the set X whenever $\mathbf{x}_1$ and $\mathbf{x}_2$ are, hence X is convex.

(2) If $f(\mathbf{x})$ is a convex function over a closed convex set X, then a local minimum of f in X is also the global minimum.

Proof The argument here is exactly as in 1.1.

(3) The set of points at which a convex function $f(\mathbf{x})$ takes on its global minimum is a convex set.

Proof If the global minimum is taken on at one point only, the result follows directly. Otherwise, let $\mathbf{x}_1^*$, $\mathbf{x}_2^*$ be points where $f(\mathbf{x}_1^*)=f(\mathbf{x}_2^*)=f^*$; then at any point on the line segment

$$\mathbf{x}=\theta\mathbf{x}_1^*+(1-\theta)\mathbf{x}_2^*, \quad 0 \leqslant \theta \leqslant 1,$$

and by convexity

$$f(\mathbf{x}) \leqslant \theta f(\mathbf{x}_1^*)+(1-\theta)f(\mathbf{x}_2^*) \leqslant f^*.$$

Since we cannot have $f(\mathbf{x})<f^*$, it must follow that

$$f(\mathbf{x})=f^*$$

at all points on this line segment. The argument can be extended to the convex hull of points $\mathbf{x}_1^*, \mathbf{x}_2^*, \ldots, \mathbf{x}_s^*$; if $f(\mathbf{x})=f^*$ at all of these, then $f(\mathbf{x})=f^*$ also at all points

$$\mathbf{x}=\sum_{r=1}^{s}\mu_r\mathbf{x}_r^* \quad \text{where } \mu_r\geqslant 0, \quad \sum_{1}^{s}\mu_r=1.$$

It follows also that there cannot be two or more strong local minima of f since a strong local minimum exists only if there is no adjacent point at which f has the same value.

(4) If $f(\mathbf{x})$ is a convex function and all $c_i(\mathbf{x})$ are concave functions, then X is a convex set; and if $\lambda_i\geqslant 0$, all i, the Lagrangian $L(\mathbf{x}, \boldsymbol{\lambda})$ is a convex function.

This follows immediately from the definitions, and it means that a set of sufficient conditions for $\mathbf{x}^*$ to be a global minimum for $f(\mathbf{x})$, $\mathbf{x}\in X$, with $f, c_i\in C^1$, are

$$\begin{aligned}
&\mathbf{x}^*\in X, c_i(\mathbf{x}^*)=0, \quad i\in I, \quad c_i(\mathbf{x}^*)\geqslant 0, \quad \text{all } i\\
&\mathbf{a}_i(\mathbf{x}^*) \text{ independent for } i\in I \qquad (1.45)\\
&\mathbf{g}(\mathbf{x}^*)-\{\mathbf{A}(\mathbf{x}^*)\}^{\mathrm{T}}\boldsymbol{\lambda}^*=0, \quad \boldsymbol{\lambda}^*\geqslant 0, \quad \{\mathbf{c}(\mathbf{x}^*)\}^{\mathrm{T}}\boldsymbol{\lambda}^*=0
\end{aligned}$$

and f convex, c_i concave. The same conditions will be sufficient for a local minimum when they hold in some neighbourhood of $\mathbf{x}^*$. They are not, as has been said, the most general – all we really need is that changes in the value of f for feasible displacements, that is those confined to the tangent plane of the active set of constraints, should be nonnegative – but they are general enough to form a basis for many practical algorithms.

The conditions (1.45) are illustrated in Fig. 1.8, which shows the situation when the Kuhn–Tucker conditions are satisfied in three very simple cases. The feasible region is shown shaded, and the arrows show the common direction of the gradient $\mathbf{g}(\mathbf{x}^*)$ of the objective function $f(\mathbf{x})$, and the gradient $\mathbf{a}(\mathbf{x}^*)$ of the (single) active constraint $c(\mathbf{x})\geqslant 0$. The sufficiency conditions do not hold in Fig. 1.8(a) and (b) – in (a) because f is not convex (confirm this) and in (b) because $c(\mathbf{x})$ is not concave (confirm this). In neither of these cases is the stationary point $\mathbf{x}^*$ a minimum, as can be verified. Figure 1.8(c) illustrates the position when conditions (1.45) hold.

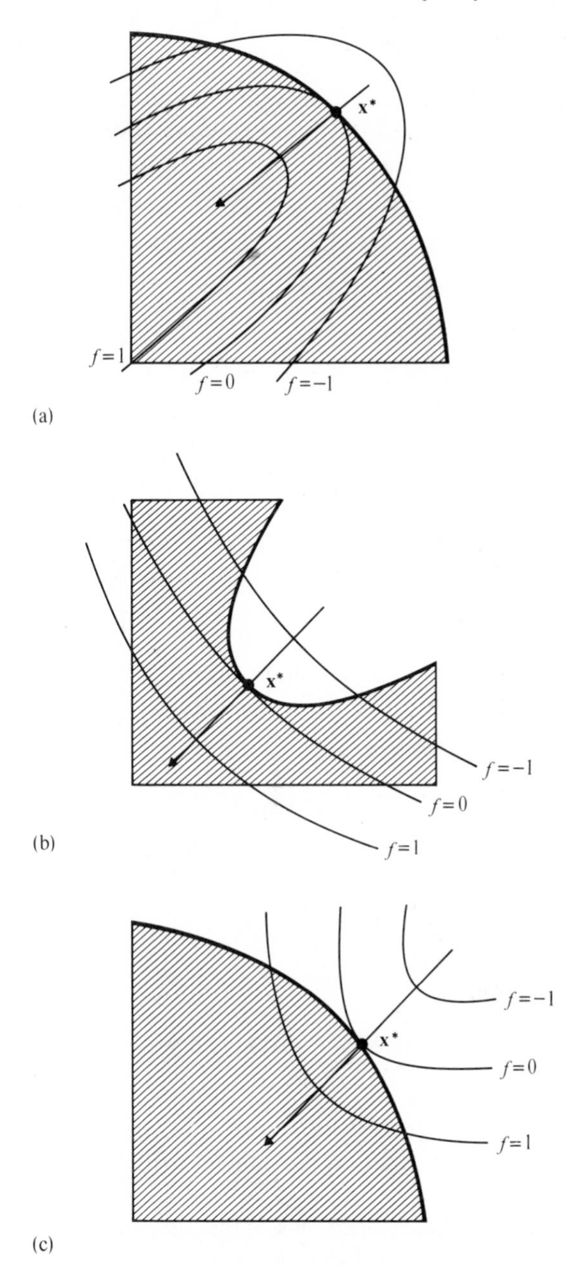

Fig. 1.8. Kuhn–Tucker conditions hold; $\mathbf{x}^*$ not minimum in (a) and (b), is in (c).

1.3 Duality

Related problems of maximising and minimising, termed dual problems, can be defined in various contexts. For example, the problem:

(A) given an acute-angled triangle PQR, find the minimum value as X ranges over the interior of the triangle of

$$f(X)=|XP|+|XQ|+|XR|$$

has as its dual the problem:
(B) for the same triangle, find the maximum side of a circumscribing equilateral triangle.

The minimising point X for problem (A) is such that the maximum equilateral triangle in (B) has sides perpendicular to PX, QX, RX (confirm this).

We now consider the special case of a function of two vector variables $\mathbf{x}$, $\mathbf{y}$, each restricted to a feasible set $\mathbf{x}\in X\subset R^n, \mathbf{y}\in Y\subset R^m$. Assume $F(\mathbf{x}, \mathbf{y})$ continuous for such $\mathbf{x}, \mathbf{y}$ and then define

$$\min_{\mathbf{x}\in X} F(\mathbf{x}, \mathbf{y}) = m(\mathbf{y}) \quad \text{and} \quad \max_{\mathbf{y}\in Y} F(\mathbf{x}, \mathbf{y}) = M(\mathbf{x}).$$

Since for any feasible $\mathbf{x}, \mathbf{y}$,

$$m(\mathbf{y}) \leqslant F(\mathbf{x}, \mathbf{y}) \leqslant M(\mathbf{x})$$

it follows that any $M(\mathbf{x})$ provides an upper bound for all the $m(\mathbf{y})$, and hence that

$$\max_{\mathbf{y}\in Y} m(\mathbf{y}) \leqslant M(\mathbf{x}), \quad \text{all } \mathbf{x}\in X$$

and similarly that

$$\min_{\mathbf{x}\in X} M(\mathbf{x}) \geqslant m(\mathbf{y}), \quad \text{all } \mathbf{y}\in Y$$

so that

$$\max_{\mathbf{y}\in Y} \min_{\mathbf{x}\in X} F(\mathbf{x}, \mathbf{y}) \leqslant \min_{\mathbf{x}\in X} \max_{\mathbf{y}\in Y} F(\mathbf{x}, \mathbf{y}). \tag{1.46}$$

Note that a special case of this is that for any two-dimensional array a_{ij}, $i=1, \ldots, m$, $j=1, \ldots, n$,

$$\max_i \min_j a_{ij} \leqslant \min_j \max_i a_{ij}. \tag{1.47}$$

The two problems

$$(A)\ \max_{\mathbf{y}\in Y}\left\{\min_{\mathbf{x}\in X} F(\mathbf{x},\mathbf{y})\right\},\quad \text{i.e. } \max_{\mathbf{y}\in Y} m(\mathbf{y})$$

$$(B)\ \min_{\mathbf{x}\in X}\left\{\max_{\mathbf{y}\in Y} F(\mathbf{x},\mathbf{y})\right\},\quad \text{i.e. } \min_{\mathbf{x}\in X} M(\mathbf{x})$$

are dual problems; if the optimum values of the variables and functions are respectively $\mathbf{y}^*$, $\mathbf{x}^*$, m^*, M^*, then from (1.46)

$$m^* \leqslant M^*. \tag{1.48}$$

The difference is called the duality gap, and clearly an interesting class of problems is formed by those for which the duality gap is zero, which turns out to be the case for the Lagrangian under quite general conditions. This is a most important result, perhaps the most important in the whole theory of constrained optimisation, and has far-reaching consequences on the development of algorithms and on the interpretation of results.

1.3.1 Saddle-point conditions and duality

We prove first the important saddle-point property of the Lagrangian, and then state and prove two forms of dual relationship. There are others, some under less restrictive conditions, but those given lead directly to the algorithms for linear and nonlinear programming discussed in later chapters.

It is worth mentioning that the concept of duality may be linked with the Legendre transformation which dualises the description of a surface as a locus of points to an envelope of tangent planes. A good treatment of this is in Aoki (1971) and it is not treated further here.

Saddle-point behaviour If

(i) f, c_i are C^1 functions of $\mathbf{x}$, and f is convex, c_i is concave
(ii) there exist $\boldsymbol{\lambda}^* \geqslant 0$ and $\mathbf{x}^* \in X$, where $X = \{\mathbf{x} \mid c_i(\mathbf{x}) \geqslant 0\}$ such that $\mathbf{g}(\mathbf{x}^*) - \{\mathbf{A}(\mathbf{x}^*)\}^{\mathrm{T}}\boldsymbol{\lambda}^* = 0$, $\{\mathbf{c}(\mathbf{x}^*)\}^{\mathrm{T}}\boldsymbol{\lambda}^* = 0$,

then

(iii) $L(\mathbf{x}^*, \boldsymbol{\lambda}) \leqslant L(\mathbf{x}^*, \boldsymbol{\lambda}^*) \leqslant L(\mathbf{x}, \boldsymbol{\lambda}^*)$ for all $\boldsymbol{\lambda} \geqslant 0$.

The property (iii) defines a saddle point of L, and the theorem in this form states that an $\mathbf{x}^*$ which minimises f subject to $\mathbf{x} \in X$,

and the corresponding multipliers $\boldsymbol{\lambda}^*$, provide saddle values for the Lagrangian.
Proof

$$L(\mathbf{x}^*, \boldsymbol{\lambda}^*) - L(\mathbf{x}^*, \boldsymbol{\lambda}) = \{\mathbf{c}(\mathbf{x}^*)\}^{\mathrm{T}}(\boldsymbol{\lambda} - \boldsymbol{\lambda}^*)$$
$$= \{\mathbf{c}(\mathbf{x}^*)\}^{\mathrm{T}}\boldsymbol{\lambda} \geqslant 0 \quad \text{(from (ii))}$$

$$L(\mathbf{x}, \boldsymbol{\lambda}^*) - L(\mathbf{x}^*, \boldsymbol{\lambda}^*) = f(\mathbf{x}) - \{\mathbf{c}(\mathbf{x})\}^{\mathrm{T}}\boldsymbol{\lambda}^*$$
$$- f(\mathbf{x}^*) + \{\mathbf{c}(\mathbf{x}^*)\}^{\mathrm{T}}\boldsymbol{\lambda}^*$$

and since f is convex, c_i concave,

$$f(\mathbf{x}) - f(\mathbf{x}^*) \geqslant (\mathbf{x} - \mathbf{x}^*)^{\mathrm{T}}\mathbf{g}(\mathbf{x}^*)$$

$$c_i(\mathbf{x}) - c_i(\mathbf{x}^*) \leqslant (\mathbf{x} - \mathbf{x}^*)^{\mathrm{T}}\mathbf{a}_i(\mathbf{x}^*).$$

Also

$$\lambda_i^* \geqslant 0$$

and so

$$L(\mathbf{x}, \boldsymbol{\lambda}^*) - L(\mathbf{x}^*, \boldsymbol{\lambda}^*) \geqslant (\mathbf{x} - \mathbf{x}^*)^{\mathrm{T}}\left[\mathbf{g}(\mathbf{x}^*) - \sum_{i=1}^{m} \lambda_i^* \mathbf{a}_i(\mathbf{x}^*)\right] = 0. \quad (1.49)$$

Conversely, if for f, c_i as in (i) we are given (iii), then (ii) follows, i.e. the saddle point property implies the Kuhn–Tucker conditions.
Proof If $L(\mathbf{x}^*, \boldsymbol{\lambda}) \leqslant L(\mathbf{x}^*, \boldsymbol{\lambda}^*)$ for all $\boldsymbol{\lambda} \geqslant 0$, it is necessary that

$$\nabla_{\lambda} L(\mathbf{x}^*, \boldsymbol{\lambda}^*) \geqslant 0, \quad \boldsymbol{\lambda}^{*\mathrm{T}} \nabla_{\lambda} L(\mathbf{x}^*, \boldsymbol{\lambda}^*) = 0;$$

that is, $c_i(\mathbf{x}^*) \geqslant 0$, $\lambda_i^* c_i(\mathbf{x}^*) = 0$, so $\mathbf{x}^* \in X$, $\{\mathbf{c}(\mathbf{x}^*)\}^{\mathrm{T}}\boldsymbol{\lambda}^* = 0$ as in (ii).
If $L(\mathbf{x}^*, \boldsymbol{\lambda}^*) \leqslant L(\mathbf{x}, \boldsymbol{\lambda}^*)$ for all $\mathbf{x}$, it is necessary that

$$\nabla_{\mathbf{x}} L(\mathbf{x}^*, \boldsymbol{\lambda}^*) = 0;$$

that is, $\mathbf{g}(\mathbf{x}^*) - \{\mathbf{A}(\mathbf{x}^*)\}^{\mathrm{T}}\boldsymbol{\lambda}^* = 0$, which completes the proof.
The behaviour of the function L is then as sketched in Fig. 1.9. Note that the critical (saddle) value of the Lagrangian is again the same as the minimum value, $f(\mathbf{x}^*)$, of f.

Dual theorem A Suppose the primal problem to be as before, minimise $f(\mathbf{x})$ subject to $\mathbf{x} \in X$, where

$$X = \{\mathbf{x} \mid c_i(\mathbf{x}) \geqslant 0, i = 1, \ldots, m\} \quad (1.50)$$

and consider the m constraints partitioned into $i \in J_1$, $i \in J_2$.

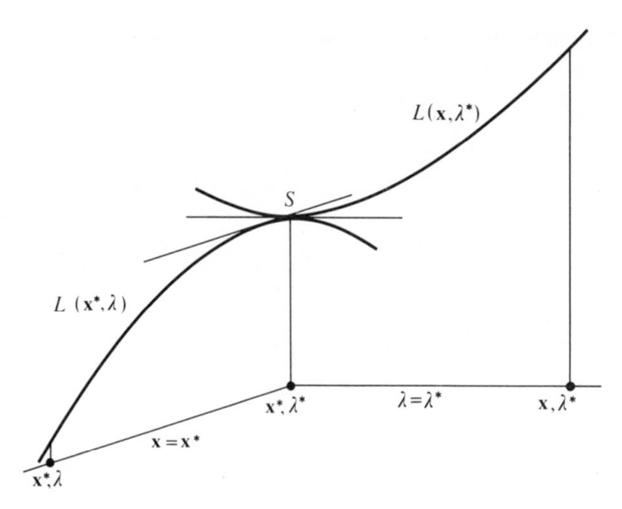

Fig. 1.9. Saddle-point property of the Lagrangian.

Define the set X_1 as $\{\mathbf{x} \mid c_i(\mathbf{x}) \geqslant 0, i \in J_1\}$ and X_2 as $\{\mathbf{x} \mid c_i(\mathbf{x}) \geqslant 0, i \in J_2\}$ with the whole solution set $X = X_1 \cap X_2$. This division of the constraints is sometimes convenient; if J_2 is empty, then X_2 is simply R^n.

Define a partial Lagrangian

$$L_1(\mathbf{x}, \bar{\boldsymbol{\lambda}}) = f(\mathbf{x}) - \sum_{i \in J_1} \bar{\lambda}_i c_i(\mathbf{x}) = f(\mathbf{x}) - \{\bar{\mathbf{c}}(\mathbf{x})\}^{\mathrm{T}} \bar{\boldsymbol{\lambda}}, \tag{1.51}$$

where $\bar{\mathbf{c}}$ contains only c_i, $i \in J_1$; and consider the dual objective function

$$\phi(\bar{\boldsymbol{\lambda}}) = \min_{\mathbf{x} \in X_2} L_1(\mathbf{x}, \bar{\boldsymbol{\lambda}}), \quad \bar{\boldsymbol{\lambda}} \geqslant 0. \tag{1.52}$$

Suppose that f is convex, c_i concave so that X_1, X_2, X are all convex sets and L_1 is a convex function of $\mathbf{x}$ for any $\bar{\boldsymbol{\lambda}} \geqslant 0$. The dual problem is defined to be

$$\max \phi(\bar{\boldsymbol{\lambda}}) \text{ subject to } \bar{\boldsymbol{\lambda}} \geqslant 0 \tag{1.53}$$

and the dual theorem states that:

if $\mathbf{x}^*, \bar{\boldsymbol{\lambda}}^*$ exist such that $\mathbf{x}^* \in X$, $\bar{\boldsymbol{\lambda}}^* \geqslant 0$, and $\{\bar{\mathbf{c}}(\mathbf{x}^*)\}^{\mathrm{T}} \bar{\boldsymbol{\lambda}}^* = 0$, and $\mathbf{x}^*$ minimises $L_1(\mathbf{x}, \bar{\boldsymbol{\lambda}}^*)$ over $\mathbf{x} \in X_2$, then $\mathbf{x}^*$ is optimal for the primal problem, $\bar{\boldsymbol{\lambda}}^*$ is optimal for the dual problem, and $f(\mathbf{x}^*) = \phi(\bar{\boldsymbol{\lambda}}^*)$.

Proof (i) We prove first that ϕ is a concave function of $\bar{\boldsymbol{\lambda}}$.

$$\phi\{(1-\theta)\bar{\boldsymbol{\lambda}}_1+\theta\bar{\boldsymbol{\lambda}}_2\}=\min_{\mathbf{x}\in X_2}[f(\mathbf{x})-(1-\theta)\{\bar{\mathbf{c}}(\mathbf{x})\}^{\mathrm{T}}\bar{\boldsymbol{\lambda}}_1-\theta\{\bar{\mathbf{c}}(\mathbf{x})\}^{\mathrm{T}}\bar{\boldsymbol{\lambda}}_2]$$

$$=\min_{\mathbf{x}\in X_2}\{(1-\theta)L_1(\mathbf{x},\bar{\boldsymbol{\lambda}}_1)+\theta L_1(\mathbf{x},\bar{\boldsymbol{\lambda}}_2)\}$$

$$\geqslant(1-\theta)\phi(\bar{\boldsymbol{\lambda}}_1)+\theta\phi(\bar{\boldsymbol{\lambda}}_2)\quad\text{for } 0<\theta<1.$$

Hence ϕ is a concave function of $\bar{\boldsymbol{\lambda}}$.

(ii) If $\mathbf{x}\in X, \bar{\boldsymbol{\lambda}}\geqslant 0$, we prove that the primal objective is not less than the dual.

$$\begin{aligned}f(\mathbf{x})&\geqslant\min_{\mathbf{x}\in X}f(\mathbf{x})\\&=f(\mathbf{x}^*)\\&\geqslant\min_{\mathbf{x}\in X}\{L_1(\mathbf{x},\bar{\boldsymbol{\lambda}})\}\quad\text{for any } \bar{\boldsymbol{\lambda}}\geqslant 0\text{, since } \{\mathbf{c}(\mathbf{x})\}^{\mathrm{T}}\bar{\boldsymbol{\lambda}}\geqslant 0\\&\geqslant\min_{\mathbf{x}\in X_2}\{L_1(\mathbf{x},\bar{\boldsymbol{\lambda}})\}\quad\text{since } X_2\supset X\\&=\phi(\bar{\boldsymbol{\lambda}}).\end{aligned}$$

Further, since $f(\mathbf{x}^*)$ is independent of $\bar{\boldsymbol{\lambda}}$ we may consider the maximising $\bar{\boldsymbol{\lambda}}$, $\bar{\boldsymbol{\lambda}}^*$, and write

$$f(\mathbf{x})\geqslant f(\mathbf{x}^*)\geqslant\phi(\bar{\boldsymbol{\lambda}}^*)\geqslant\phi(\bar{\boldsymbol{\lambda}}). \tag{1.54}$$

(iii) For any $\mathbf{x}\in X$, and $\bar{\boldsymbol{\lambda}}^*$ satisfying given conditions,

$$\begin{aligned}f(\mathbf{x})&\geqslant f(\mathbf{x})-\{\bar{\mathbf{c}}(\mathbf{x})\}^{\mathrm{T}}\bar{\boldsymbol{\lambda}}^*\\&>\min_{\mathbf{x}\in X_2}L_1(\mathbf{x},\bar{\boldsymbol{\lambda}}^*)\\&=L_1(\mathbf{x}^*,\bar{\boldsymbol{\lambda}}^*)\quad\text{since } \mathbf{x}^* \text{ minimises } L_1(\mathbf{x},\bar{\boldsymbol{\lambda}}^*)\text{ over } \mathbf{x}\in X_2\\&=f(\mathbf{x}^*)\quad\text{since } \{\bar{\mathbf{c}}(\mathbf{x}^*)\}^{\mathrm{T}}\bar{\boldsymbol{\lambda}}^*=0.\end{aligned}$$

So $\mathbf{x}^*$ is optimal for the primal. Also

$$\begin{aligned}\phi(\bar{\boldsymbol{\lambda}}^*)&=\min_{\mathbf{x}\in X_2}L_1(\mathbf{x},\bar{\boldsymbol{\lambda}}^*)\\&=L_1(\mathbf{x}^*,\bar{\boldsymbol{\lambda}}^*)\quad\text{by definition}\\&=f(\mathbf{x}^*).\end{aligned}$$

Hence from (1.54) $\bar{\boldsymbol{\lambda}}^*$ is optimal for the dual.

This result is the basis of algorithms which obtain estimates for the primal problem by using information about Lagrange multipliers, as described in Chapter 5.

Duality theorem B – symmetric form This form of dual relation suggests the main duality theorem of linear programming, though, as will be seen, the proof used here cannot be applied to the linear case which has to be proved separately (see Chapter 3).
Theorem For any function $K(\mathbf{x}, \boldsymbol{\lambda})$, differentiable of class C^2, convex in $\mathbf{x} \geqslant 0$, concave in $\boldsymbol{\lambda} \geqslant 0$, we can define two problems.

$$\text{Primal:} \quad \min_{S_1} M_1(\mathbf{x}, \boldsymbol{\lambda}) = K(\mathbf{x}, \boldsymbol{\lambda}) - \boldsymbol{\lambda}^{\mathrm{T}} \nabla_\lambda K$$

$$\text{where } S_1 = \{\mathbf{x}, \boldsymbol{\lambda} \mid \mathbf{x} \geqslant 0, \boldsymbol{\lambda} \geqslant 0, \nabla_\lambda K \leqslant 0\} \qquad (1.55)$$

$$\text{Dual:} \quad \max_{S_2} M_2(\mathbf{x}, \boldsymbol{\lambda}) = K(\mathbf{x}, \boldsymbol{\lambda}) - \mathbf{x}^{\mathrm{T}} \nabla_{\mathbf{x}} K$$

$$\text{where } S_2 = \{\mathbf{x}, \boldsymbol{\lambda} \mid \mathbf{x} \geqslant 0, \boldsymbol{\lambda} \geqslant 0, \nabla_{\mathbf{x}} K \geqslant 0\}. \qquad (1.56)$$

Then the theorem states that

$$\text{(i)} \quad \min_{S_1} M_1 \geqslant \max_{S_2} M_2 \qquad (1.57)$$

and (ii) assuming the Kuhn–Tucker conditions hold for both problems, then if there exist $\mathbf{x}^*, \boldsymbol{\lambda}^*$ which solve one, and if $\nabla_{\mathbf{x}}^2 K^*$ is positive definite, $\nabla_\lambda^2 K^*$ is negative definite, then $\mathbf{x}^*, \boldsymbol{\lambda}^*$ also solve the other and the optimal values for both problems are equal.
Proof From convexity in $\mathbf{x}$, concavity in $\boldsymbol{\lambda}$, we have for $\mathbf{x}_1$, $\boldsymbol{\lambda}_1 \in S_1$ and $\mathbf{x}_2, \boldsymbol{\lambda}_2 \in S_2$

$$K(\mathbf{x}_1, \boldsymbol{\lambda}_1) - K(\mathbf{x}_2, \boldsymbol{\lambda}_2) \geqslant (\mathbf{x}_1 - \mathbf{x}_2)^{\mathrm{T}} \nabla_{\mathbf{x}} K(\mathbf{x}_2, \boldsymbol{\lambda}_2) + (\boldsymbol{\lambda}_1 - \boldsymbol{\lambda}_2)^{\mathrm{T}} \nabla_{\boldsymbol{\lambda}} K(\mathbf{x}_1, \boldsymbol{\lambda}_1)$$

and so

$$\begin{aligned} M_1(\mathbf{x}_1, \boldsymbol{\lambda}_1) - M_2(\mathbf{x}_2, \boldsymbol{\lambda}_2) &\geqslant \mathbf{x}_1^{\mathrm{T}} \nabla_{\mathbf{x}} K(\mathbf{x}_2, \boldsymbol{\lambda}_2) - \boldsymbol{\lambda}_2^{\mathrm{T}} \nabla_{\boldsymbol{\lambda}} K(\mathbf{x}_1, \boldsymbol{\lambda}_1) \\ &\geqslant 0 \text{ by the definitions of } S_1 \text{ and } S_2. \end{aligned}$$

Relation (i) then follows.

To prove (ii), we assume the existence of $\mathbf{x}^*, \boldsymbol{\lambda}^*$ which maximise M_2 in S_2, and show that they also lie in S_1, and that $M_1 = M_2$.

Introduce multipliers $\mathbf{u}$ and write a Lagrangian for M_2,

$$\begin{aligned} L_2(\mathbf{x}, \boldsymbol{\lambda}, \mathbf{u}) &= M_2(\mathbf{x}, \boldsymbol{\lambda}) + \mathbf{u}^{\mathrm{T}} \nabla_{\mathbf{x}} K(\mathbf{x}, \boldsymbol{\lambda}) \\ &= K(\mathbf{x}, \boldsymbol{\lambda}) + (\mathbf{u} - \mathbf{x})^{\mathrm{T}} \nabla_{\mathbf{x}} K(\mathbf{x}, \boldsymbol{\lambda}) \end{aligned} \qquad (1.58)$$

Using the Kuhn–Tucker conditions means that at the maximising point $\mathbf{x}^*, \boldsymbol{\lambda}^*$ subject to $\mathbf{x}, \boldsymbol{\lambda} \in S_2$

$$\nabla_{\mathbf{x}} L_2^* \leqslant 0, \quad \nabla_{\boldsymbol{\lambda}} L_2^* \leqslant 0, \quad \mathbf{x}^{*\mathrm{T}} \nabla_{\mathbf{x}} L_2^* + \boldsymbol{\lambda}^{*\mathrm{T}} \nabla_{\boldsymbol{\lambda}} L_2^* = 0 \tag{1.59}$$

and

$$\nabla_{\mathbf{u}} L_2^* \geqslant 0, \quad \mathbf{u}^{*\mathrm{T}} \nabla_{\mathbf{u}} L_2^* = 0, \quad \mathbf{u}^* \geqslant 0. \tag{1.60}$$

Substituting from (1.58) and simplifying, conditions (1.59) reduce to

$$\nabla_{\mathbf{x}}^2 K^* (\mathbf{u}^* - \mathbf{x}^*) \leqslant 0, \quad \nabla_{\boldsymbol{\lambda}} K^* + \{\nabla_{\boldsymbol{\lambda}} \nabla_{\mathbf{x}} K^*\}(\mathbf{u}^* - \mathbf{x}^*) \leqslant 0 \tag{1.61}$$

and

$$\mathbf{x}^{*\mathrm{T}} \{\nabla_{\mathbf{x}}^2 K^*\}(\mathbf{u}^* - \mathbf{x}^*) + \boldsymbol{\lambda}^{*\mathrm{T}} [\nabla_{\boldsymbol{\lambda}} K^* + \{\nabla_{\boldsymbol{\lambda}} \nabla_{\mathbf{x}} K^*\}(\mathbf{u}^* - \mathbf{x}^*)] = 0. \tag{1.62}$$

A sum of elements all nonpositive can only be zero if each individual element is zero and hence from (1.61), (1.62)

$$\mathbf{x}^{*\mathrm{T}} \{\nabla_{\mathbf{x}}^2 K^*\}(\mathbf{u}^* - \mathbf{x}^*) = 0.$$

Since also $\mathbf{u}^{*\mathrm{T}} \{\nabla_{\mathbf{x}}^2 K^*\}(\mathbf{u}^* - \mathbf{x}^*) \leqslant 0$, then $(\mathbf{u}^{*\mathrm{T}} - \mathbf{x}^{*\mathrm{T}})\{\nabla_{\mathbf{x}}^2 K^*\}(\mathbf{u}^* - \mathbf{x}^*) \leqslant 0$. But it is stated that $\nabla_{\mathbf{x}}^2 K^*$ is positive definite and hence

$$\mathbf{u}^* = \mathbf{x}^*. \tag{1.63}$$

(Note that this is the step which cannot be applied to the case when K is linear since then $\nabla_{\mathbf{x}}^2 K^*$ is identically zero.)

From (1.61), (1.63)

$$\nabla_{\boldsymbol{\lambda}} K^* \leqslant 0 \quad \text{so } \mathbf{x}^*, \boldsymbol{\lambda}^* \in S_1.$$

From (1.62), (1.63)

$$\boldsymbol{\lambda}^{*\mathrm{T}} \nabla_{\boldsymbol{\lambda}} K^* = 0.$$

So

$$M_1(\mathbf{x}^*, \boldsymbol{\lambda}^*) = K(\mathbf{x}^*, \boldsymbol{\lambda}^*).$$

And substituting from (1.58), (1.63), in (1.60) gives

$$\mathbf{x}^{*\mathrm{T}} \nabla_{\mathbf{x}} K^* = 0 \quad \text{so } M_2(\mathbf{x}^*, \boldsymbol{\lambda}^*) = K(\mathbf{x}^*, \boldsymbol{\lambda}^*)$$

which proves the theorem.

1.4 Summary of basic results

Since these results are the foundation of the methods described in the following chapters, it will be useful to collect together the main ones.

Unconstrained variables At minimum of $f(\mathbf{x}), f \in C^1$, a necessary condition is $\mathbf{g}(\mathbf{x}^*)=0$; sufficient conditions are $\mathbf{g}(\mathbf{x}^*)=0$, f locally convex or, if $f \in C^2$, $\mathbf{G}(\mathbf{x}^*)$ positive definite.

Equality constraints At minimum of $f(\mathbf{x})$, subject to $\mathbf{c}(\mathbf{x})=0$, with $f, c_i \in C^1$, necessary conditions (for a normal point) are

$$\mathbf{c}(\mathbf{x}^*)=0, \quad \mathbf{g}(\mathbf{x}^*)=\sum_{i=1}^{m} \lambda_i^* \mathbf{a}_i(\mathbf{x}^*).$$

These are also necessary conditions for a critical point at $\mathbf{x}^*, \boldsymbol{\lambda}^*$ of the Lagrangian

$$L(\mathbf{x}, \boldsymbol{\lambda})=f(\mathbf{x})-\sum_{i=1}^{m} \lambda_i c_i(\mathbf{x}).$$

Sufficient conditions are, in addition to the above, that $L(\mathbf{x}, \boldsymbol{\lambda}^*)$ is a locally convex function of $\mathbf{x}$.

If the i-th constraint is $c_i(\mathbf{x})=b_i$, then $\partial f^*/\partial b_i\,|_{b_i=0}=\lambda_i^*$. The Lagrangian $L(\mathbf{x}^*, \boldsymbol{\lambda}^*)=f(\mathbf{x}^*)$.

Inequality constraints At minimum of $f(\mathbf{x})$, subject to $\mathbf{c}(\mathbf{x}) \geqslant 0$, with $f, c_i \in C^1$, necessary conditions (for a normal point) are

$$\mathbf{c}(\mathbf{x}^*) \geqslant 0, \quad \mathbf{g}(\mathbf{x}^*)=\sum_{i=1}^{m} \lambda_i^* \mathbf{a}_i(\mathbf{x}^*), \quad \lambda_i^* c_i(\mathbf{x}^*)=0, \quad \lambda_i^* \geqslant 0.$$

These are also necessary conditions for $\mathbf{x}^*, \boldsymbol{\lambda}^*$ to be a saddle point of the Lagrangian, such that

$$L(\mathbf{x}^*, \boldsymbol{\lambda}) \leqslant L(\mathbf{x}^*, \boldsymbol{\lambda}^*) \leqslant L(\mathbf{x}, \boldsymbol{\lambda}^*) \quad \text{for } \boldsymbol{\lambda} \geqslant 0, \text{ all } \mathbf{x}.$$

Sufficient conditions are, in addition to the above, that f is convex, c_i are concave so that L is locally convex. For an active constraint, $c_i(\mathbf{x}) \geqslant b_i$, $\partial f^*/\partial b_i\,|_{b_i=0}=\lambda_i^*$; for an inactive constraint, $\lambda_i^*=0$. Again $L(\mathbf{x}^*, \boldsymbol{\lambda}^*)=f(\mathbf{x}^*)$.

Notes

1.1 Cauchy–Schwartz inequality

For any two vectors $\mathbf{x}, \mathbf{y}$

$$(a\mathbf{x}+\mathbf{y})^{\mathrm{T}}(a\mathbf{x}+\mathbf{y})>0$$

and so

$$a^2(\mathbf{x}^T\mathbf{x})+2a(\mathbf{x}^T\mathbf{y})+(\mathbf{y}^T\mathbf{y})>0$$

This holds for all a, and so

$$(\mathbf{x}^T\mathbf{y})^2 \leqslant (\mathbf{x}^T\mathbf{x})(\mathbf{y}^T\mathbf{y})$$

as required.

1.2

A vector $\mathbf{h}$ is said to be an admissible direction at $\mathbf{x}$ if there exists a sequence $\mathbf{x}_r$, $r=1, 2, \ldots$, in X and a sequence of numbers $\varepsilon_r>0$ such that

$$\lim \mathbf{x}_r=\mathbf{x}, \quad \lim (\mathbf{x}_r-\mathbf{x})/\varepsilon_r=\mathbf{h};$$

the tangent cone of X at $\mathbf{x}$ is the set of admissible directions at $\mathbf{x}$. Then if $\mathbf{x}^*$ is the minimum of $f(\mathbf{x})$ for $\mathbf{x}\in X$, it follows that $\mathbf{h}^T\mathbf{g}(\mathbf{x}^*)\geqslant 0$ for all admissible $\mathbf{h}$. Also if $X=\{\mathbf{x} \mid c_i(\mathbf{x})\geqslant 0\}$, then any admissible $\mathbf{h}$ is contained in the set $\{\mathbf{h} \mid \mathbf{h}^T\mathbf{a}_i(\mathbf{x}^*)\geqslant 0\}$. However it is possible for members of this set not to be admissible – as in 1.2.4. If the members are all admissible, that is if the two sets are identical, then the *constraint qualification* is satisfied.

This is so if the $\mathbf{a}_i(\mathbf{x}^*)$ are independent for $i\in I$, the active set at $\mathbf{x}^*$. For let $\mathbf{h}$ satisfy $\mathbf{h}^T\mathbf{a}_i(\mathbf{x}^*)\geqslant 0$, $i\in I$, and let $\mathbf{h}^*$ satisfy $\mathbf{h}^{*T}\mathbf{a}_i(\mathbf{x}^*)=0$ for $i\in I_p$, a subset of I. If I_p contains p elements, then by the Implicit Function Theorem there exist differentiable functions $X_1, X_2, \ldots, X_p$ such that

$$x_i=X_i(x_{p+1}, x_{p+2}, \ldots, x_n) \quad \text{for } i\in I_p$$

(taking I_p as the set $i=\{1, 2, \ldots, p\}$)

and

$$c_i(X_1, X_2, \ldots, X_p, x_{p+1}, x_{p+2}, \ldots, x_n)=0$$

for all $x_{p+1}, x_{p+2}, \ldots, x_n$ in some neighbourhood of $\mathbf{x}^*$. Then we can show that the points

$$\mathbf{x}_r=\mathbf{x}_r(x_{p+1}, x_{p+2}, \ldots, x_n),$$

with

$$x_j=x_j^*+\varepsilon_r(h_j+dh_j^*), \quad j=p+1, p+2, \ldots, n,$$

where $\varepsilon_r>0$, $\lim_{r\to\infty}\varepsilon_r=0$, and $d>0$, satisfy

$$\lim_{r\to\infty} \mathbf{x}_r=\mathbf{x}^*, \quad \lim_{r\to\infty} (\mathbf{x}_r-\mathbf{x})/\varepsilon_r=\mathbf{h}+d\mathbf{h}^*.$$

Also $c_i(\mathbf{x}_r)>0$ for sufficiently large r, so that $\mathbf{x}_r \in X$; and so $\mathbf{h}+d\mathbf{h}^*$ is an admissible direction. Now d can be made arbitrarily small, and so the constraint qualification is satisfied under these conditions.

Problems

(1) Using first-order necessary conditions, find stationary points of the function

$$f(\mathbf{x})=x_1^4+4x_1^2x_2^2-2x_1^2+2x_2^2-1.$$

Determine their nature using second-order conditions.

(2) Find the stationary point of the function

$$f(\mathbf{x})=2x_1^2+x_1x_2+x_2^2+x_2x_3+x_3^2-6x_1-7x_2-8x_3+9$$

and verify that it is the global minimum point.

(3) (a) Show that the function

$$f(\mathbf{x})=(x_1^2+x_2^2+x_3^2)^2$$

has a strong minimum at $(0, 0, 0)$.

(b) Show that the Hessian $\mathbf{G}(\mathbf{x})$ is positive semi-definite only at $(0, 0, 0)$.

(c) Suggest sufficient conditions for $\mathbf{x}^*$ to be a strong minimum when $\mathbf{G}(\mathbf{x}^*)$ is not positive definite.

(4) Prove that the function $1/x$ is strictly convex for $x>0$ and strictly concave for $x<0$.

(5) Prove that if $f_i(\mathbf{x})$, $i=1, 2, \ldots, m$, are all convex functions, then

(i) $\sum \alpha_i f_i(\mathbf{x})$ is convex if $\alpha_i>0$, all i.

(ii) $f(\mathbf{x})=\sup_i f_i(\mathbf{x})$ is convex on the region where it is finite.

(6) Find the stationary point $\mathbf{x}^*$ of

$$f(\mathbf{x})=x_1^2+x_2^2+x_3^2$$

subject to

$$c_1(\mathbf{x})\equiv x_1+x_2+3x_3-2=0,$$

$$c_1(\mathbf{x})\equiv 5x_1+2x_2+x_3-5=0,$$

and determine its nature. Verify that $\mathbf{g}(\mathbf{x}^*)$ can be written as a linear combination of $\mathbf{a}_1(\mathbf{x}^*), \mathbf{a}_2(\mathbf{x}^*)$.

(7) Show that $\mathbf{x}^T = (1, 1, 1)$ is a stationary point of

$$f(\mathbf{x}) = x_1^2 + 2x_2^2 + 4x_3^2 + 5x_1x_2$$

subject to

$$c_1(\mathbf{x}) \equiv x_1 + x_2^2 + 3x_2x_3 - 5 = 0,$$
$$c_2(\mathbf{x}) \equiv x_1^2 + 5x_1x_2 + 3x_3^2 - 9 = 0,$$

and determine its nature.

(8) Verify that the function

$$f(\mathbf{x}) = x_1^2 + x_2^2 + x_3^2$$

subject to

$$c_1(\mathbf{x}) \equiv x_1^2 - 4x_2 = 0,$$
$$c_2(\mathbf{x}) \equiv x_1 - x_2 - 1 = 0$$

has a minimum at $\mathbf{x}^{*T} = (2, 1, 0)$, but that $\mathbf{g}(\mathbf{x}^*)$ is not expressible as $\lambda_1\mathbf{a}_1(\mathbf{x}^*) + \lambda_2\mathbf{a}_2(\mathbf{x}^*)$.

(9) Write down the Kuhn–Tucker conditions for the problem

$$\text{minimise } f(\mathbf{x}) \equiv -x_1^3 + x_2^2 - 2x_1x_3^2 \quad \text{subject to}$$
$$c_1(\mathbf{x}) \equiv 2x_1 + x_2^2 + x_3 - 5 = 0$$
$$c_2(\mathbf{x}) \equiv 5x_1^2 - x_2^2 - x_3 \geqslant 2$$
$$x_1 \geqslant 0, \quad x_2 \geqslant 0, \quad x_3 \geqslant 0.$$

and verify that they are satisfied at $(1, 0, 3)$.

(10) Write down the Kuhn–Tucker conditions for the problem

$$\text{minimise } f(\mathbf{x}) \equiv x_1^2 + x_2^2 + x_3^2 \quad \text{subject to}$$
$$c_1(\mathbf{x}) \equiv -x_1 + x_2 - x_3 \geqslant -10,$$
$$c_2(\mathbf{x}) \equiv x_1 + x_2 + 4x_3 \geqslant 20,$$

and obtain the solution in any manner.

(11) In the theory of consumer behaviour it is assumed that each individual has a utility function $U(\mathbf{x})$, where $\mathbf{x}$ is a vector of consumption, i.e. the quantities of n specified commodities. He seeks to maximise this, subject to a budget constraint $\mathbf{p}^T\mathbf{x} \leqslant d$, where $\mathbf{p}$ is a price vector, and d is a (scalar) limit on spending. Write down the conditions for optimum $\mathbf{x}$, assuming that $U(\mathbf{x}) \in C^1$. Show that the solution is unique when U is concave

everywhere, and that the vector $\mathbf{x}$ can be expressed explicitly in terms of $\mathbf{p}$ and d when U is quadratic. What is the economic interpretation of the Lagrange multiplier?

(12) $f(\mathbf{x})$ is a convex function of $\mathbf{x}$, and $c_i(\mathbf{x})$ are concave functions of $\mathbf{x}$, $i=1,\ldots,m$. Prove that the function $f(\mathbf{x})-\sum_{i=1}^{m}\lambda_i c_i(\mathbf{x})$, $\lambda_i \geqslant 0$, is convex. If the function $\phi(\boldsymbol{\lambda})$ is defined by

$$\phi(\boldsymbol{\lambda})=\min_{\mathbf{x}\in X}\left\{f(\mathbf{x})-\sum_{i=1}^{m}\lambda_i c_i(\mathbf{x})\right\},$$

where X is the feasible set $c_i(\mathbf{x})\geqslant 0$, prove that ϕ is a concave function of $\boldsymbol{\lambda}$.

(13) Given that $f(\mathbf{x})$ is the quadratic function

$$f(\mathbf{x})=\tfrac{1}{2}\mathbf{x}^{\mathrm{T}}\mathbf{G}\mathbf{x},$$

where $\mathbf{G}$ is a positive definite symmetric matrix, having distinct eigenvalues $\lambda_1>\lambda_2>\cdots>\lambda_n$ and corresponding eigenvectors $\mathbf{y}_1,\mathbf{y}_2,\ldots,\mathbf{y}_n$, verify that
(a) $\max f(\mathbf{x})$ subject to $\tfrac{1}{2}\mathbf{x}^{\mathrm{T}}\mathbf{x}=1$ is λ_1, and that, for $K\leqslant n$,
(b) $\max f(\mathbf{x})$ subject to $\tfrac{1}{2}\mathbf{x}^{\mathrm{T}}\mathbf{x}=1$, $\mathbf{x}^{\mathrm{T}}\mathbf{y}_i=0$, $i=1,2,\ldots,K-1$, is λ_K.

(14) Prove that if $f(\mathbf{x})$ is a convex function, then the problem

$$\text{minimise } f(\mathbf{x}) \text{ subject to } \mathbf{x}\geqslant 0$$

has a dual problem

$$\text{maximise } h(\mathbf{y})=f(\mathbf{y})-\mathbf{y}^{\mathrm{T}}\mathbf{g}(\mathbf{y})$$

subject to $\mathbf{y}\geqslant 0, \mathbf{g}(\mathbf{y})\geqslant 0$, where $\mathbf{g}(\mathbf{x})=\nabla f(\mathbf{x})$. Show that if both problems have solutions $\mathbf{x}^*, \mathbf{y}^*$ then $f(\mathbf{x}^*)=h(\mathbf{y}^*)$.

(15) Prove that if $\mathbf{x}^*$ is the solution to the problem

$$\text{minimise } f(\mathbf{x})=\mathbf{c}^{\mathrm{T}}\mathbf{x}+\tfrac{1}{2}\mathbf{x}^{\mathrm{T}}\mathbf{G}\mathbf{x}$$

subject to $\mathbf{A}\mathbf{x}=\mathbf{b}, \mathbf{x}\geqslant 0$, then the problem:

$$\text{maximise } h(\boldsymbol{\lambda})=\boldsymbol{\lambda}^{\mathrm{T}}\mathbf{b}$$

subject to $\mathbf{A}^{\mathrm{T}}\boldsymbol{\lambda}\leqslant\mathbf{G}\mathbf{x}^*+\mathbf{c}$ has a solution $\boldsymbol{\lambda}^*$, and that

$$\mathbf{x}^{*\mathrm{T}}\mathbf{G}\mathbf{x}^*=\boldsymbol{\lambda}^{*\mathrm{T}}\mathbf{b}-\mathbf{c}^{\mathrm{T}}\mathbf{x}^*.$$

(16) Given that the set X is convex and nonempty, and if the system

$$f_i(\mathbf{x})<0, \quad h_j(\mathbf{x})=0, \quad i=1,2,\ldots,m, j=1,\ldots,n,$$

where the $f_i(\mathbf{x})$ are convex and the $h_j(\mathbf{x})$ are linear, has no solution in X, prove that there exist $\boldsymbol{\lambda}, \boldsymbol{\mu}$ such that

$$\sum_{i=1}^{m} \lambda_i f_i(\mathbf{x}) + \sum_{j=1}^{n} \mu_j h_j(\mathbf{x}) \geq 0$$

for all $\mathbf{x} \in X$, where $\boldsymbol{\lambda} \geq 0$, with same $\lambda_i > 0$ or $\mu_j \neq 0$, or both.

(17) If $X, Y \subset R^n$ are convex sets with no interior point in common, prove that there exists a hyperplane $\mathbf{c}^{\mathrm{T}}\mathbf{x} = f^*$ which separates X and Y so that $\mathbf{c}^{\mathrm{T}}\mathbf{x} \geq f^*$ for all $\mathbf{x} \in X$ and $\mathbf{c}^{\mathrm{T}}\mathbf{y} \leq f^*$ for all $\mathbf{y} \in Y$. (Consider the set

$$U = \{\mathbf{x} - \mathbf{y}, \mid \mathbf{x} \in X, \mathbf{y} \in Y\}.$$

Show that U is convex, and that the point 0 either lies on the boundary of U or is not in U.)

2

Unconstrained optimisation

It is important to be able to find the minimum of $f(\mathbf{x})$ for unbounded $\mathbf{x}$ since this is often part of the solution process even when the problem is one for which $\mathbf{x}$ is restricted. In particular many methods require line searches, that is finding the minimum of $f(\mathbf{x})$ for points $\mathbf{x}$ along a line $\mathbf{x}=\mathbf{x}_r+\alpha\mathbf{p}_r$ determined by given $\mathbf{x}_r$, $\mathbf{p}_r$.

In Section 2.1 methods for line searches are described, based on dissection – Fibonacci, golden section – or using fitted polynomials. In Section 2.2 there is a brief description of general search methods, that is methods for minimising $f(\mathbf{x})$ which use only values of f at a sequence of chosen points. Section 2.3 then introduces gradient methods, steepest descents and conjugate gradients. In all these cases the solution is obtained by a step-by-step procedure, the directions of successive steps being related by vector equations. Powell's method is mentioned here also since, although it does not use gradients explicitly, it is based on the conjugate-gradient property. Most of the methods in this and the next section have the property of quadratic termination and so an account of the properties of quadratic functions is included. Finally Section 2.4 describes the methods which update step directions by multiplying by matrices, motivated originally by Newton's method and so often called quasi Newton; the best known of these is the Davidon–Fletcher–Powell or variable metric method.

2.1 Line-search methods

2.1.1 Dissection methods

(a) Fibonacci search Suppose that the function f of one variable is known to have a minimum, at x^*, and to be unimodal; this is a weaker property than convexity, and is defined by the conditions

$$f(x_1)>f(x_2)>f(x^*) \quad \text{for } x_1, x_2 \text{ such that } x_1<x_2<x^* \text{ or } x^*<x_2<x_1.$$

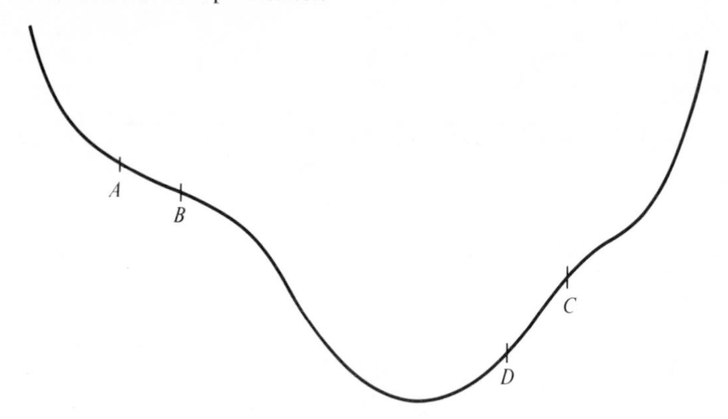

Fig. 2.1. Unimodal function.

A unimodal function is shown in Fig. 2.1. If its values at three points A, B, C, B being between A and C, are known to satisfy $f_B < f_A$, $f_B < f_C$, then clearly the minimum x^* must lie between A and C.

With information about values of f at other points within AC, the interval of uncertainty can be reduced and we may ask what are the best set of points to use if we want to produce the smallest final interval while calculating a fixed number, N, of extra values. An even distribution of N points would give a final interval reduced by the factor $2/(N+1)$, but a much more efficient process is available in the Fibonacci sequence.

Consider the insertion of one extra point D within the interval AC in Fig. 2.1; if D is in AB, then according as $f_D < f_B$ or $> f_B$ the new uncertainty interval is AB or DC, while if D is in BC then it is BC or AD. If $AB = S$, $BC = l$, $l > S$, we can ensure a new interval l by putting D in BC and making $AD = BC$, and any other arrangement could result in an interval larger than l. The process will then be repeated starting with the appropriate three points, ABD or BDC.

Now consider the final (N-th) point. This should be inserted so that it and the existing interior point trisect the interval, that is

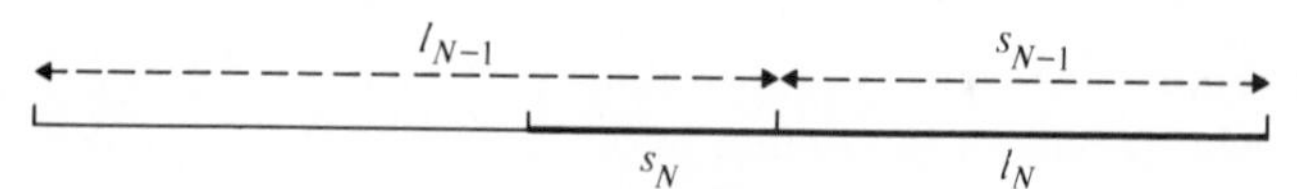

Fig. 2.2. Fibonacci search.

the best arrangement will be to have $l_N = 2S_N = I_N$. Then at the previous stage (Fig. 2.2)

$$l_{N-1} = l_N + S_N \quad \text{and} \quad S_{N-1} = l_N,$$

and so

$$l_{N-2} = l_{N-1} + S_{N-1} = l_{N-1} + l_N.$$

This is the recurrence relation for the Fibonacci sequence $\{I_r\}$,

$$1, 2, 3, 5, 8, 13, 21, \ldots$$

where each term is the sum of the two previous ones, and $I_0 = 1$, $I_1 = 2$. It follows that if N points are inserted within an interval according to the procedure:

(1) find f at P_1, P_2, points symmetrically placed at I_{N-1}/I_N of the interval from the two end points;

(2) pick out the new interval within which x^* must lie;

(3) insert the next new point symmetrically in this interval with respect to the interior point already there and find f;

(4) repeat 2 and 3 until N points have been inserted,

then this process will give a reduction factor of I_1/I_N in the uncertainty interval.

Example 2.1 For example, consider finding the minimum of the function $f(x) = (3 - 4x)/(1 + x^2)$ within the range $1 \leqslant x \leqslant 6.5$, using 8 points. Since I_8 is 55, this will locate x^* to ± 0.1. The first two points are inserted at (34/55) of the interval from the ends, that is at 3.1, 4.4, giving $f(3.1) = -0.886$, $f(4.4) = -0.717$ while $f(1) = -0.5$, $f(6.5) = -0.532$. Then the search process will proceed using $f(2.3) = -0.986$, $f(1.8) = -0.991$, $f(1.5) = -0.923$, $f(2.0) = -1$, $f(2.1) = -0.886$, $f(1.9) = -0.998$ with the final interval $1.9 \leqslant x^* \leqslant 2.1$. The actual minimum in this case is at $x = 2$.

(b) Golden section Solving the difference equation for the Fibonacci sequence gives

$$I_r = A\left(\frac{1+\sqrt{5}}{2}\right)^r + B\left(\frac{1-\sqrt{5}}{2}\right)^r,$$

and so the ratio I_{N-1}/I_N tends for large N to $2/(1+\sqrt{5}) = 0.618$, the "golden section", so called by the ancient Greeks. The name comes from its link with an "ideal" rectangle, for which

$$\text{short side} : \text{long side} = \text{long side} : \text{diagonal}$$

This common ratio l is then given by $l^2=0.618$. An alternative search procedure is then to insert the first two points at 0.618 of the interval from the ends, and subsequently to insert each new point symmetrically as above. This method is independent of N, but it does not have the trisection property within the final interval.

Example 2.2 Minimise the same function as in Example 2.1, using the golden section method.

Now the first two points are at 3.101, 4.399 and the eighth point is at 1.924, with the adjacent points 1.803, 1.990 and 2.111. Of course in this case the process could be continued if greater accuracy were desired, the interval of uncertainty being reduced by a factor of 0.618 each time.

2.1.2 Fitted polynomial methods

(a) Fitted quadratic Values of f at any N points can be fitted by an interpolating $(N-1)$-th order polynomial; the simplest polynomial which can have a minimum is a quadratic, so $N \geqslant 3$. From 3 values f_1, f_2, f_3 at x_1, x_2, x_3 it is easy to show that the point x_M at which the interpolating quadratic $y(x)$ has a stationary value is given by

$$x_M = \frac{1}{2}\frac{(x_2^2-x_3^2)f_1+(x_3^2-x_1^2)f_2+(x_1^2-x_2^2)f_3}{(x_2-x_3)f_1+(x_3-x_1)f_2+(x_1-x_2)f_3} \tag{2.1}$$

or, an alternative form which may be easier to calculate,

$$x_M = \tfrac{1}{4}(x_1+2x_2+x_3)-\tfrac{1}{4}(x_3-x_1)\times\frac{\{(f_2-f_1)/(x_2-x_1)+(f_3-f_2)/(x_3-x_2)\}}{\{(f_3-f_2)/(x_3-x_2)-(f_2-f_1)/(x_2-x_1)\}} \tag{2.2}$$

Here x_1, x_2, x_3 are any points, not necessarily in order of magnitude. If they are equally spaced, so that $x_2-x_1=x_3-x_2=h$, (2.2) becomes

$$x_M = x_2-\tfrac{1}{2}h(f_3-f_1)/(f_1-2f_2+f_3) \tag{2.3}$$

The (constant) second derivative y'' is

$$Q=-2\frac{\{(x_2-x_3)f_1+(x_3-x_1)f_2+(x_1-x_2)f_3\}}{(x_1-x_2)(x_2-x_3)(x_3-x_1)} \tag{2.4}$$

which reduces in the equally spaced case to the second difference approximation

$$Q=\delta^2/h^2=(f_1-2f_2+f_3)/h^2 \tag{2.5}$$

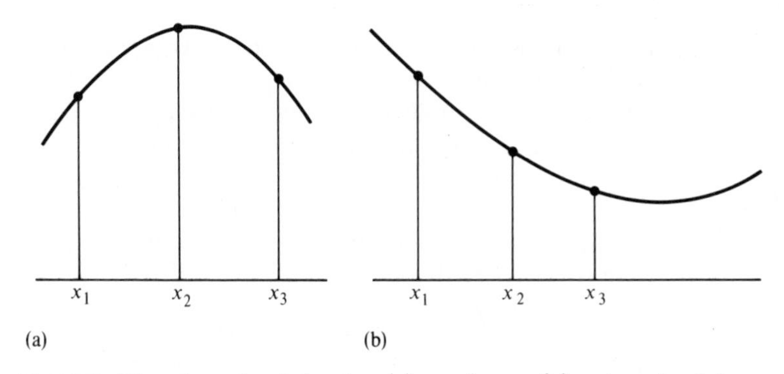

Fig. 2.3. Fitted quadratic having (a) maximum (b) external minimum.

We would like to choose 3 points, derive from them an x_M, and set up an iterative scheme whereby x_M replaces one of the original points and the process is repeated. However, we may find $Q<0$, in which case x_M is a maximum, or x_M may be an extrapolated point outside the interval; these cases are shown in Fig. 2.3, and either would spoil the process.

Safeguards against either of these eventualities have to be incorporated in a general process. However, if we start with 3 points as with the dissection methods such that $x_1<x_2<x_3$, $f_2<f_1, f_2<f_3$, then it can be seen that $Q>0$ and that $x_1<x_M<x_3$. If we take the appropriate set of 3 new points, then the process will converge to the minimum x^*. A variant is to calculate Q and x_M from x_1, x_2, x_3 and subsequently to regard Q as an estimate of the new second derivative and to use the two outer points of the new interval only, and then

$$x_M = -\tfrac{1}{2}(f_1 - f_2)/\{Q(x_1 - x_2)\} + \tfrac{1}{2}(x_1 + x_2) \tag{2.6}$$

is the point where the quadratic $y(x)$ through (x_1, f_1) and (x_2, f_2) with $y''=Q$ has its minimum value. Q may need to be updated during the calculation.

Example 2.3 Minimise the same function as in Examples 2.1, 2.2 by this method.

Take starting values

$$x_1=1, \qquad x_2=3.75 \qquad x_3=6.5$$
$$f_1=-0.5 \quad f_2=-0.797 \quad f_3=-0.532$$

From (2.3), $x_M=2.828$, $Q=0.0743$, $f_M=-0.924$.

Then choosing $x_1 = 1$, $x_2 = 2.828$ and $x_3 = 3.75$ as the new points produces from (2.2) $x_M = 2.780$.

Continuing the process produces the following set of values

$$2.828, 2.780, 2.492, 2.382, 2.259, 2.192, 2.129, 2.102, \ldots$$

Use of (2.6) with the initial Q gives the values 3.102, 2.640, but then diverges so the value for Q must be improved. This method was first produced by Powell (1964).

(b) Fitted cubic A cubic can be fitted to values of f and f' at 2 points x_1, x_2, and its stationary value found. More safeguards are needed since a cubic need not have a stationary point, or will have both a minimum and a maximum. Davidon's (1959) method implements this procedure starting with a point x_1 where $f_1' < 0$, and first finding a point $x_2 > x_1$ where either $f_2' > 0$ or $f_2 > f_1$; the cubic must then have a minimum between x_1 and x_2. An outline of this method is developed in Problem 3 at the end of this chapter.

Fitted polynomial methods, since they use information on the shape of the function, are generally more efficient than simple search though also somewhat more complicated to implement.

2.2 General search methods

Methods which minimise $f(\mathbf{x})$ using function values only are specially valuable in applications where the objective function is non-differentiable or has discontinuous first derivatives; for functions with a degree of smoothness other methods are generally more efficient. A brief description only is given here since these methods are essentially heuristic and more dependent on skill and ingenuity in programming than on mathematical formulation.

(a) Pattern search Evaluation of f at all points of a regular grid would certainly reveal the minimum but at prohibitive cost; the obvious requirement is to direct attention to the region of space where the minimum is and to restrict its position more and more closely as with Fibonacci line search. Hooke and Jeeves (1961) produced a simple and widely used method which examines successively how f changes when steps $\pm h_i$ are taken from a base point $\mathbf{x}_1$ in the direction $\mathbf{e}_i$, the unit vector along the i-th coordinate axis. If either step decreases f, you adopt that point as the new start for steps along $\mathbf{e}_{i+1}$. The set of explorations along $\mathbf{e}_i$, $i=1,2,\ldots n$, constitute an exploratory move, and if the base point $\mathbf{x}_1$ has been changed to $\mathbf{x}_2$, then the direction $\mathbf{x}_2-\mathbf{x}_1$ is a favourable one and a new base point $2\mathbf{x}_2-\mathbf{x}_1$ is taken as the origin of a new exploration. If this new search provides a point $\mathbf{x}_3$ with a value of f less than the value at $\mathbf{x}_2$, then $\mathbf{x}_3-\mathbf{x}_2$ is regarded as a favourable direction and the new search starts from $2\mathbf{x}_3-\mathbf{x}_2$; otherwise the starting point is $\mathbf{x}_2$. If the base point does not change, then all steps h_i are halved and the exploration repeated.

(b) Simplex method An alternative (Nelder and Mead 1965) is to consider points at the vertices of a geometrical figure, initially a simplex, that is a figure with equally spaced vertices. If values are $f_0, f_1, f_2, \ldots, f_n$, with $f_0<f_1<\cdots<f_n$, then the worst vertex $\mathbf{x}_n$ is removed and replaced by its reflection in the centroid of the others, so that

$$\mathbf{x}_{n+1}=\frac{2}{n}\left(\sum_{0}^{n-1}\mathbf{x}_i\right)-\mathbf{x}_n.$$

If $f_{n+1}<f_0$ this is a favourable direction and $\mathbf{x}_{n+1}$ is moved further out in the direction $(1/n)\sum\mathbf{x}_i-\mathbf{x}_n$. If $f_0<f_{n+1}<f_{n-1}$, the process is repeated, using the vertices with $\mathbf{x}_{n+1}$ replacing $\mathbf{x}_n$. If $f_{n+1}>f_n$ the point $\mathbf{x}_{n+1}$ is rejected and the initial figure is contracted.

2.3 Gradient methods

In all the processes now considered, we start with a point $\mathbf{x}_1$ and proceed to the minimum $\mathbf{x}^*$, by a succession of steps; at each step the direction changes, so that the r-th step is given by

$$\mathbf{x}_{r+1}=\mathbf{x}_r+\alpha_r\mathbf{p}_r,$$

where α_r, $\mathbf{p}_r$ are defined in terms of $\mathbf{g}_r$, the gradient vector defined by

$$\mathbf{g}_r^{\mathrm{T}}=\left(\frac{\partial f}{\partial x_1}, \frac{\partial f}{\partial x_2}, \ldots, \frac{\partial f}{\partial x_n}\right)$$

at $\mathbf{x}_r$, and other quantities. For the methods in this section the relations involve only products of vectors, whereas in Section 2.4, they involve products of matrices and vectors and so require rather more storage space and time per step.

2.3.1 Steepest descents

The simplest procedure is to take $\mathbf{p}_r = -\mathbf{g}_r$, that is at each step to proceed along the direction in which (as shown in Section 1.1.2) f is locally decreasing at the fastest rate, the direction of steepest descent. The minimum along this direction, found by a line search, is then used as the start of the next step. This process is defined by

$$\begin{aligned} \mathbf{x}_{r+1} &= \mathbf{x}_r - \alpha_r \mathbf{g}_r, \\ \mathbf{g}_{r+1}^{\mathrm{T}} \mathbf{g}_r &= 0, \end{aligned} \tag{2.7}$$

the second relation being a direct consequence of the minimum property of $\mathbf{x}_{r+1}$. (Alternatively and even more simply we can take $\alpha_r = \text{constant}$.)

These methods can be proved to converge to a local minimum under quite general conditions but convergence can be very slow (see the note in Section 2.4.1). The behaviour of the method is entirely dependent on the scaling of the independent variables. If we write

$$X_i = \lambda_i x_i,$$

then

$$\frac{\partial f}{\partial X_i} = \frac{1}{\lambda_i} \frac{\partial f}{\partial x_i},$$

and so the direction of $\mathbf{g}$ can be altered arbitrarily by changing $\boldsymbol{\lambda}$. Since there is no obvious way of choosing $\boldsymbol{\lambda}$ ab initio, the method is basically unsatisfactory.

Example 2.4 Evaluate the first 3 points in minimising

$$f(\mathbf{x}) = 9x_1^2 - 8x_1x_2 + 3x_2^2$$

by steepest descents starting at $\mathbf{x}_1^{\mathrm{T}} = (1, 1)$.

$$\{\mathbf{g}(\mathbf{x})\}^{\mathrm{T}} = (18x_1 - 8x_2, -8x_1 + 6x_2),$$

so $\mathbf{g}_1^{\mathrm{T}} = (10, -2)$ and the first step is along $-\mathbf{g}_1$.

$$\mathbf{x}_2^{\mathrm{T}} = (1 - 10\alpha_1, 1 + 2\alpha_1),$$

where

$$\alpha_1 = (10^2 + 2^2)/(18 \cdot 10^2 + 2 \cdot 8 \cdot 10 \cdot 2 + 6 \cdot 2^2) = 0.048\,51$$

so

$$\mathbf{x}_2^{\mathrm{T}} = (0.515, 1.097), \quad \mathbf{g}_2^{\mathrm{T}} = (0.492, 2.462), \quad \alpha_2 = 0.2953$$

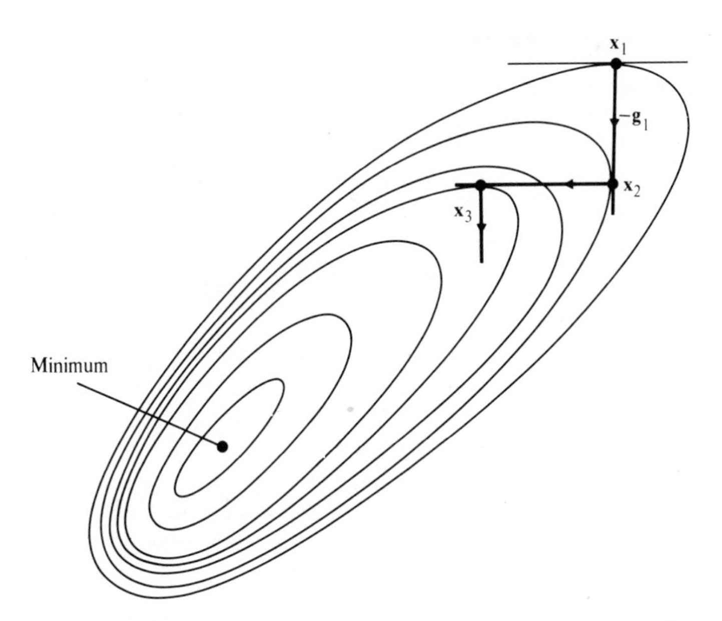

Fig. 2.4. Method of steepest descents for function of two variables.

(since $\mathbf{g}_2 \perp \mathbf{g}_1$ we don't really need to work it out in this simple case, we could just write $\mathbf{g}_2^{\mathrm{T}} = (2, 10)$ and use the corresponding α_2 which would be 0.0726)

$$\mathbf{x}_3^{\mathrm{T}} = (0.370, 0.370),$$

and similarly

$$\mathbf{x}_4^{\mathrm{T}} = (0.191, 0.406).$$

The process has reduced f from $f_1 = 4$ via $f_2 = 1.478$, $f_3 = 0.548$, to $f_4 = 0.202$.

However, if the axes are rotated to the principal axes by the transformation

$$\begin{bmatrix} x_1 \\ x_2 \end{bmatrix} = \begin{bmatrix} 1/\sqrt{5} & -2/\sqrt{5} \\ 2/\sqrt{5} & 1/\sqrt{5} \end{bmatrix} \begin{bmatrix} y_1 \\ y_2 \end{bmatrix},$$

then $f(\mathbf{y}) = y_1^2 + 11y_2^2$, and if then $Y_1 = y_1$, $Y_2 = \sqrt{11}y_2$, we have

$$f(\mathbf{Y}) = Y_1^2 + Y_2^2$$

and clearly for this function the method of steepest descents produces the minimum in one step!

2.3.2 Gradient properties of quadratic functions

A quadratic function

$$f(\mathbf{x})=\tfrac{1}{2}\mathbf{x}^{\mathrm{T}}\mathbf{G}\mathbf{x}-\mathbf{b}^{\mathrm{T}}\mathbf{x}, \tag{2.8}$$

where $\mathbf{G}$ is a positive definite symmetric matrix, is the simplest function which has a minimum. Any efficient minimisation algorithm must be able to find this minimum, and a method with *quadratic termination* finds it in a finite number of steps. Such methods are based on the algebraic properties of quadratic functions and so a brief account of these is given first; of course the methods are designed to minimise general functions and their efficiency depends on choosing the appropriate forms for quantities which are equivalent for quadratics.

(a) General properties of minima along fixed directions For $f(\mathbf{x})$ defined by (2.8), with

$$\mathbf{g}_r=\mathbf{G}\mathbf{x}_r-\mathbf{b} \tag{2.9}$$

$$\mathbf{g}_r-\mathbf{g}_s=\mathbf{G}(\mathbf{x}_r-\mathbf{x}_s) \tag{2.10}$$

The minimum of f along the direction $\mathbf{p}_r$ is at $\mathbf{x}_{r+1}=\mathbf{x}_r+\alpha_r\mathbf{p}_r$, where from (2.7)

$$\mathbf{g}_{r+1}^{\mathrm{T}}\mathbf{p}_r=0 \tag{2.11}$$

and so, using (2.9),

$$\alpha_r=-\mathbf{p}_r^{\mathrm{T}}\mathbf{g}_r/\mathbf{p}_r^{\mathrm{T}}\mathbf{G}\mathbf{p}_r, \tag{2.12}$$

and using (2.10)

$$\mathbf{g}_{r+1}=\mathbf{g}_r+\alpha_r\mathbf{G}\mathbf{p}_r \tag{2.13}$$

and the change in f,

$$f_{r+1}-f_r=-\tfrac{1}{2}(\mathbf{p}_r^{\mathrm{T}}\mathbf{g}_r)^2/\mathbf{p}_r^{\mathrm{T}}\mathbf{G}\mathbf{p}_r. \tag{2.14}$$

It follows that the function f is strictly decreasing for $\mathbf{p}_r^{\mathrm{T}}\mathbf{g}_r\neq 0$, since $\mathbf{G}$ is positive definite. The overall minimum $\mathbf{x}^*$ is where

$$\mathbf{g}^*=\mathbf{G}\mathbf{x}^*-\mathbf{b}=0. \tag{2.15}$$

(b) Conjugate directions A set of directions $\mathbf{p}_r$ which satisfy the condition

$$\mathbf{p}_r^{\mathrm{T}}\mathbf{G}\mathbf{p}_s=0 \quad \text{for} \quad r\neq s \tag{2.16}$$

are known as conjugate directions with respect to **G**. We can list certain properties of these

(1) They are linearly independent: for the relation $\sum \lambda_i \mathbf{p}_i = 0$ implies, from (2.16), $\lambda_i \mathbf{p}_i^T \mathbf{G} \mathbf{p}_i = 0$, all i, and $\mathbf{p}_i^T \mathbf{G} \mathbf{p}_i \neq 0$ since **G** is positive definite.

(2) A set of **p**'s can be constructed from any set of n linearly dependent **u**'s by a Gram–Schmidt type procedure

$$\mathbf{p}_1 = \mathbf{u}_1$$

$$\mathbf{p}_{r+1} = \mathbf{u}_{r+1} + \sum_{i=1}^{r} \beta_i^{(r)} \mathbf{p}_i, \tag{2.17}$$

where

$$\beta_i^{(r)} = -(\mathbf{u}_{r+1}^T \mathbf{G} \mathbf{p}_i)/(\mathbf{p}_i^T \mathbf{G} \mathbf{p}_i).$$

The relation (2.16) can be verified by induction for this set, starting with $\mathbf{p}_1$, $\mathbf{p}_2$. Because of (1) there are at most n nonzero **p**'s.

(3) Any other vector can be written in terms of a basis set of **p**'s. In particular the vector from the initial point $\mathbf{x}_1$ to the minimum $\mathbf{x}^*$ is

$$\mathbf{x}^* - \mathbf{x}_1 = \sum_1^n \alpha_r \mathbf{p}_r, \quad \text{where } \alpha_r = \mathbf{p}_r^T \mathbf{G}(\mathbf{x}^* - \mathbf{x}_1)/\mathbf{p}_r^T \mathbf{G} \mathbf{p}_r, \tag{2.18}$$

and using (2.9) and (2.15), this gives

$$\alpha_r = -\mathbf{p}_r^T \mathbf{g}_1 / \mathbf{p}_r^T \mathbf{G} \mathbf{p}_r. \tag{2.19}$$

Hence $\mathbf{x}^*$ can be reached in $k \leq n$ steps from $\mathbf{x}_1$, where the r-th step is

$$\mathbf{x}_{r+1} = \mathbf{x}_r + \alpha_r \mathbf{p}_r$$

and α_r is given by (2.19). Also, since

$$\mathbf{x}_r = \mathbf{x}_1 + \sum_{i=1}^{r-1} \alpha_i \mathbf{p}_i,$$

then using (2.10),

$$\mathbf{g}_r = \mathbf{g}_1 + \sum_{i=1}^{r-1} \alpha_i \mathbf{G} \mathbf{p}_i \tag{2.20}$$

and so

$$\mathbf{p}_r^T \mathbf{g}_1 = \mathbf{p}_r^T \mathbf{g}_r \quad \text{since } \mathbf{p}_r^T \mathbf{G} \mathbf{p}_i = 0 \quad \text{for } i \leq r-1. \tag{2.21}$$

Hence the expressions (2.12) and (2.19) are equal for a quadratic function and the steps from $\mathbf{x}_1$ to $\mathbf{x}^*$ proceed via points which are minima along each direction. Also since, for $s<r$, from (2.20)

$$\mathbf{p}_s^{\mathrm{T}}\mathbf{g}_r=\mathbf{p}_s^{\mathrm{T}}\mathbf{g}_1+\alpha_s\mathbf{p}_s^{\mathrm{T}}\mathbf{G}\mathbf{p}_s=0 \quad \text{(from (2.19))}, \tag{2.22}$$

it follows that $\mathbf{g}_r$ is orthogonal to all previous search directions and so that each point $\mathbf{x}_{r+1}$ is a minimum not only along the direction $\mathbf{p}_r$ but also with respect to the whole subspace spanned by $\mathbf{p}_1, \mathbf{p}_2, \ldots, \mathbf{p}_r$.

(4) The absolute value of the determinant $M(\mathbf{p})$ having as columns a set of directions $\mathbf{p}_r$, normalised by $\mathbf{p}_r^{\mathrm{T}}\mathbf{G}\mathbf{p}_r=1$, all r, takes its maximum value when the $\mathbf{p}_r$'s are mutually conjugate. For consider the Lagrange form

$$\Phi(\mathbf{p},\boldsymbol{\lambda})=M(\mathbf{p})+\sum_{r=1}^{n}\lambda_r(\mathbf{p}_r^{\mathrm{T}}\mathbf{G}\mathbf{p}_r-1). \tag{2.23}$$

The conditions for a stationary value of M subject to the normalising equations are $\partial\Phi/\partial p_{r,j}=0$, where $p_{r,j}$ is the j-th component of $\mathbf{p}_r$. These become

$$P_{r,j}+2\lambda_r[\mathbf{G}\mathbf{p}_r]_j=0,$$

where $P_{r,j}$ is the cofactor of $p_{r,j}$ in $M(\mathbf{p})$, and so multiplying by $p_{r,j}$ and summing over j gives

$$M+2\lambda_r=0, \quad \text{all } r,$$

so that $\lambda_r\neq 0$ since $M=0$ is certainly not the maximum; while multiplying by $p_{s,j}$, $s\neq r$, and summing over j gives

$$0+2\lambda_r\mathbf{p}_s^{\mathrm{T}}\mathbf{G}\mathbf{p}_r=0;$$

hence the $\mathbf{p}$'s must be conjugate. The usual property that if the columns of a determinant are normalised so that $\mathbf{x}_r^{\mathrm{T}}\mathbf{x}_r=1$, then the determinant has its extreme value when the columns are orthogonal follows from this by putting $\mathbf{G}=\mathbf{I}$.

2.3.3 The method of conjugate gradients

This method, due to Fletcher and Reeves (1964), uses a set of $\mathbf{p}$'s based on the negative gradients $-\mathbf{g}_r$ at successive points. We prove first that this is valid by showing that the $\mathbf{g}$'s are linearly independent – in fact that they are orthogonal; and second that the expression (2.17) takes a particularly simple form in this case.

Start with

$$\mathbf{p}_1 = -\mathbf{g}_1.$$

Then from (2.11) $\mathbf{g}_2^T\mathbf{p}_1 = 0$; hence $\mathbf{g}_2^T\mathbf{g}_1 = 0$.

The mutual orthogonality of all the $\mathbf{g}$'s is now proved by induction. Assume that $\mathbf{g}_1, \mathbf{g}_2, \ldots, \mathbf{g}_r$ are orthogonal, and that $\mathbf{p}_1, \mathbf{p}_2, \ldots, \mathbf{p}_r$ have been found from them using (2.17), so that these $\mathbf{p}$'s are conjugate. Hence from (2.22) $\mathbf{g}_{r+1}$, the gradient at the minimum point $\mathbf{x}_{r+1}$ along the search direction $\mathbf{p}_r$, is orthogonal to all previous $\mathbf{p}$, that is

$$\mathbf{g}_{r+1}^T\mathbf{p}_s = 0 \quad \text{for } 1 \leqslant s \leqslant r$$

But (2.17) gives

$$\mathbf{p}_s = -\mathbf{g}_s + \sum_{i=1}^{s-1} \beta_i^{(s-1)}\mathbf{p}_i \quad \text{for } 1 \leqslant s \leqslant r,$$

and so

$$\mathbf{g}_{r+1}^T\mathbf{g}_s = -\mathbf{g}_{r+1}^T\mathbf{p}_s + \sum_{i=1}^{s-1} \beta_i^{(s-1)}\mathbf{g}_{r+1}^T\mathbf{p}_i = 0 \qquad (2.24)$$

Hence if $\mathbf{g}_1, \mathbf{g}_2, \ldots, \mathbf{g}_r$ are orthogonal, and conjugate directions $\mathbf{p}_1$ to $\mathbf{p}_r$ are generated from them, then $\mathbf{g}_{r+1}$ is orthogonal to all previous $\mathbf{g}$'s; and since $\mathbf{g}_2^T\mathbf{g}_1 = 0$ has been shown, it follows that $\mathbf{g}_1$ to $\mathbf{g}_n$ are independent and that $\mathbf{p}_1$ to $\mathbf{p}_n$ are conjugate.

Further, (2.17) in this case becomes

$$\beta_i^{(r)} = (\mathbf{g}_{r+1}^T\mathbf{G}\mathbf{p}_i)/(\mathbf{p}_i^T\mathbf{G}\mathbf{p}_i)$$

and, from (2.13), this can be written

$$(\mathbf{p}_i^T\mathbf{G}\mathbf{p}_i)\beta_i^{(r)} = \frac{1}{\alpha_i}\mathbf{g}_{r+1}^T(\mathbf{g}_{i+1} - \mathbf{g}_i).$$

Then from (2.24),

$$\beta_i^{(r)} = 0 \quad \text{for } i < r, \qquad (2.25)$$

and from (2.12)

$$\beta_r^{(r)} = -(\mathbf{g}_{r+1}^T\mathbf{g}_{r+1})/(\mathbf{p}_r^T\mathbf{g}_r) \qquad (2.26)$$

or, using (2.22) and (2.17),

$$\beta_r^{(r)} = (\mathbf{g}_{r+1}^T\mathbf{g}_{r+1})/(\mathbf{g}_r^T\mathbf{g}_r). \qquad (2.27)$$

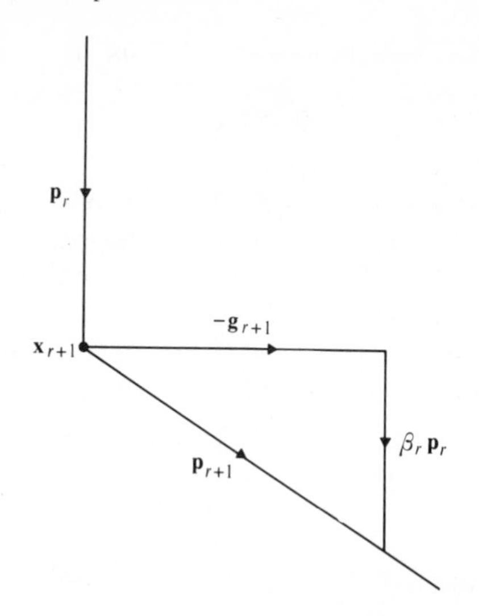

Fig. 2.5. Conjugate gradient directions.

Hence, dropping the redundant *r*, the conjugate-gradient algorithm can be summarised by the following relations;

(1) $\mathbf{p}_1 = -\mathbf{g}_1$;
(2) $\mathbf{x}_{r+1}$ is the minimum of f along the line $\mathbf{x}_r + \alpha\mathbf{p}_r$;
(3) $\mathbf{p}_{r+1} = -\mathbf{g}_{r+1} + \beta_r\mathbf{p}_r$;
(4) (i) $\beta_r = -(\mathbf{g}_{r+1}^{\mathrm{T}}\mathbf{g}_{r+1})/(\mathbf{p}_r^{\mathrm{T}}\mathbf{g}_r)$
(ii) $\beta_r = (\mathbf{g}_{r+1}^{\mathrm{T}}\mathbf{g}_{r+1})/(\mathbf{g}_r^{\mathrm{T}}\mathbf{g}_r)$
(iii) $\beta_r = \mathbf{g}_{r+1}^{\mathrm{T}}(\mathbf{g}_{r+1} - \mathbf{g}_r)/\{\mathbf{p}_r^{\mathrm{T}}(\mathbf{g}_{r+1} - \mathbf{g}_r)\}$,

where the bracketed expressions are equivalent for a quadratic function but differ for a general function, 4(ii) being the form originally suggested by Fletcher and 4(iii) that given by Polak. For a general f the algorithm does not terminate so cycles of n points are calculated, restarted each time with the current gradient.

Example 2.5 As an example of a non-quadratic function, consider

$$f(x_1, x_2) = (x_2 - x_1^2)^2 + (1 - x_1)^2$$

which has a minimum value of 0 at $x_1 = x_2 = 1$. Then

$$g_1(x_1, x_2) = -4x_1(x_2 - x_1^2) - 2 + 2x_1, \quad g_2(x_1, x_2) = 2(x_2 - x_1^2).$$

So starting at the origin

$$\mathbf{x}_1^{\mathrm{T}} = (0, 0), \quad f_1 = 1, \quad \mathbf{g}_1^{\mathrm{T}} = (-2, 0), \quad \mathbf{p}_1^{\mathrm{T}} = (2, 0)$$

and minimising $f(\mathbf{x}_1+\alpha\mathbf{p}_1)$ gives $\mathbf{x}_2^{\mathrm{T}}=(0.5898, 0)$, $f_2=0.2893$, $\mathbf{g}_2^{\mathrm{T}}=(0, -0.6957)$, so that $\beta_1=\|\mathbf{g}_2\|^2/\|\mathbf{g}_1\|^2=0.1210$ and $\mathbf{p}_2^{\mathrm{T}}=(0.2420, 0.6957)$. Minimising $f(\mathbf{x}_2+\alpha\mathbf{p}_2)$ gives

$$\mathbf{x}_3^{\mathrm{T}}=(0.8560, 0.7653), \quad f_3=0.0221$$

and the process is then restarted with $\mathbf{x}_1=\mathbf{x}_3$ and $\mathbf{p}_1=-\mathbf{g}_3$.

2.3.4 Powell's method

Methods using some of the advantages of conjugate directions without needing to calculate derivatives of f are based on the following result: if $\mathbf{x}_r$, $\mathbf{x}_s$ are separate points, and $\mathbf{p}$ a direction not parallel to $\mathbf{x}_r-\mathbf{x}_s$, and if $\mathbf{X}_r$ is the point which minimises $f(\mathbf{x}_r+\alpha\mathbf{p})$, and $\mathbf{X}_s$ the point which minimises $f(\mathbf{x}_s+\alpha\mathbf{p})$, then when f is quadratic $\mathbf{X}_r-\mathbf{X}_s$ is conjugate to $\mathbf{p}$. For since $\mathbf{X}_r$, $\mathbf{X}_s$ are minimum points

$$\mathbf{p}^{\mathrm{T}}\mathbf{g}(\mathbf{X}_r)=0, \quad \mathbf{p}^{\mathrm{T}}\mathbf{g}(\mathbf{X}_s)=0,$$

and so when f is quadratic, from (2.8), (2.9)

$$\mathbf{p}^{\mathrm{T}}\mathbf{G}(\mathbf{X}_r-\mathbf{X}_s)=0,$$

and so $(\mathbf{X}_r-\mathbf{X}_s)$ is conjugate to $\mathbf{p}$ since $\|\mathbf{X}_r-\mathbf{X}_s\|\neq 0$.

This extends immediately to the following result: if $\mathbf{x}_r$, $\mathbf{x}_s$ are separate points, and $\mathbf{p}_1, \mathbf{p}_2, \ldots, \mathbf{p}_K$ are mutually conjugate directions, and $\mathbf{x}_r-\mathbf{x}_s$ is linearly independent of the $\mathbf{p}$'s, then if $\mathbf{X}_r$, $\mathbf{X}_s$ are points minimising $f(\mathbf{x}_r+\sum_{i=1}^{K}\alpha_i\mathbf{p}_i)$, $f(\mathbf{x}_s+\sum_{i=1}^{K}\alpha_i\mathbf{p}_i)$ respectively, then $\mathbf{X}_r-\mathbf{X}_s$ is conjugate to all the $\mathbf{p}$'s.

Powell's method is an efficient algorithm using this property. It proceeds by specifying n linearly independent search directions

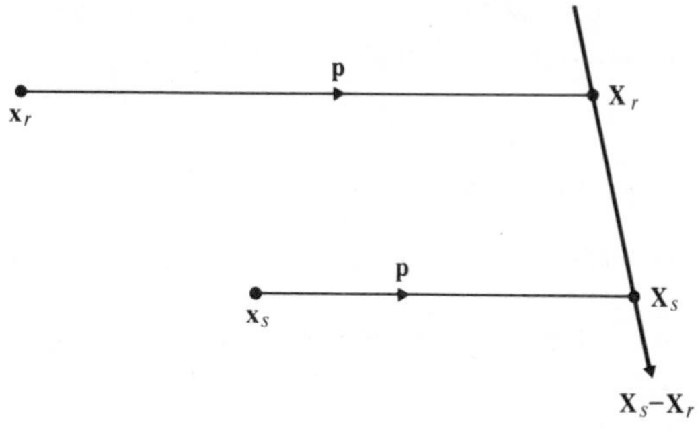

Fig. 2.6. Join of minima gives direction conjugate to **p**.

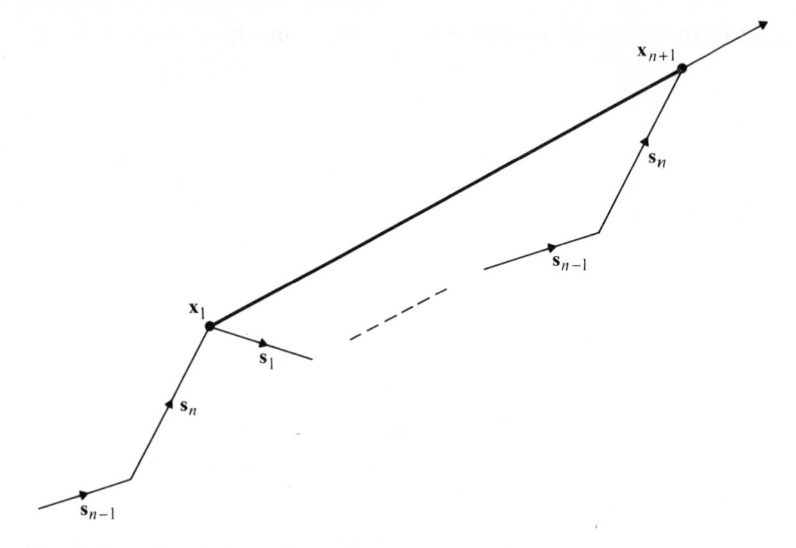

Fig. 2.7. Join of endpoints gives direction conjugate to $\mathbf{s}_n, \mathbf{s}_{n-1}, \ldots$.

and changing these if the change increases the value of the determinant $M(\mathbf{p})$; in 2.3.2(4) it was shown that this takes its maximum value when the $\mathbf{p}$'s are mutually conjugate.

In each cycle, carry out successive searches along n independent directions $\mathbf{s}_r$, starting at $\mathbf{x}_1$, with final point $\mathbf{x}_{n+1}$. Then the direction $\mathbf{x}_{n+1}-\mathbf{x}_1$ is, for f quadratic, conjugate to $\mathbf{s}_n$, if $\mathbf{x}_1$ was also found by minimising along $\mathbf{s}_n$ in the previous cycle. If $\mathbf{x}_1$ was found by minimising along a set of consecutive conjugate directions $\mathbf{s}_{K+1}, \ldots, \mathbf{s}_n$, and $\mathbf{x}_{n+1}$ in this cycle uses the same set, then $\mathbf{x}_{n+1}-\mathbf{x}_1$ is conjugate to all directions in this set.

A test is carried out first to assess the suitability of this direction using an additional calculated point at $2\mathbf{x}_{n+1}-\mathbf{x}_1$: let the values of f be

$$f(\mathbf{x}_1)=\phi_1, \quad f(\mathbf{x}_{n+1})=\phi_2, \quad f(2\mathbf{x}_{n+1}-\mathbf{x}_1)=\phi_3, \quad \text{where } \phi_1>\phi_2$$

For simplicity write $\boldsymbol{\xi}_1=\mathbf{x}_1$, $\boldsymbol{\xi}_2=\mathbf{x}_{n+1}$, $\boldsymbol{\xi}_3=2\mathbf{x}_{n+1}-\mathbf{x}_1$. These values are now used to predict the effect on the determinant $M(\mathbf{s})$ of replacing one of the $\mathbf{s}_r$ by the new direction. $M(\mathbf{s})$ has columns normalised by

$$\mathbf{s}_r^{\mathrm{T}}\mathbf{G}\mathbf{s}_r=1 \tag{2.28}$$

and from (2.12), (2.14) when $\mathbf{x}_{r+1}$ is the point minimising f along $\mathbf{x}=\mathbf{x}_r+\alpha\mathbf{s}_r$, then

$$\mathbf{x}_{r+1}=\mathbf{x}_r+\alpha_r\mathbf{s}_r, \quad f_r-f_{r+1}=\tfrac{1}{2}\alpha_r^2, \tag{2.29}$$

and so

$$\mathbf{x}_{n+1}-\mathbf{x}_1=\sum_1^n \alpha_r\mathbf{s}_r=\mu\boldsymbol{\sigma}, \tag{2.30}$$

where

$$\boldsymbol{\sigma}^{\mathrm{T}}\mathbf{G}\boldsymbol{\sigma}=1. \tag{2.31}$$

Thus replacing the column $\mathbf{s}_r$ by the new normalised vector $\boldsymbol{\sigma}$ will multiply the determinant M, from (2.30), by α_r/μ. Hence the change should be made if any $\alpha_r>\mu$ and the largest α_r should be selected, i.e. from (2.29), the direction producing the greatest change in f. Write

$$\delta=\max_r\,(f_r-f_{r+1}), \quad \text{then } \max\alpha_r=\sqrt{2}\,\delta^{\frac{1}{2}} \tag{2.32}$$

and the condition for replacing $\mathbf{s}_r$ by $\boldsymbol{\sigma}$ is

$$\mu<\sqrt{2}\,\delta^{\frac{1}{2}}. \tag{2.33}$$

To estimate μ in the same way we need to estimate the minimum of ϕ along $\boldsymbol{\sigma}$. Using the fitted quadratic approximation based on the equally spaced points $\boldsymbol{\xi}_1$, $\boldsymbol{\xi}_2$, $\boldsymbol{\xi}_3$ gives a stationary value of ϕ along $\boldsymbol{\sigma}$ as

$$\phi_3=\phi_2-\tfrac{1}{8}(\phi_1-\phi_3)^2/(\phi_1-2\phi_2+\phi_3). \tag{2.34}$$

(i) If the second difference $\phi_1-2\phi_2+\phi_3<0$, the fitted quadratic has a maximum. This condition implies $\phi_3<\phi_2<\phi_1$.

(ii) If $\phi_1-2\phi_2+\phi_3>0$, then either (a) $\phi_3>\phi_1$ or (b) $\phi_3<\phi_1$.

(a) $\phi_3>\phi_1$, so that $\boldsymbol{\xi}_s$ lies between $\boldsymbol{\xi}_1$ and $\boldsymbol{\xi}_2$.

In this case using (2.29)

$$\begin{aligned}\|\boldsymbol{\xi}_s-\boldsymbol{\xi}_1\|&=\mu_1\sigma, \quad \mu_1=\sqrt{2}(\phi_1-\phi_s)^{\frac{1}{2}},\\ \|\boldsymbol{\xi}_2-\boldsymbol{\xi}_s\|&=\mu_2\sigma, \quad \mu_2=\sqrt{2}(\phi_2-\phi_s)^{\frac{1}{2}}.\end{aligned} \tag{2.35}$$

But $\|\boldsymbol{\xi}_2-\boldsymbol{\xi}_1\|=\mu\sigma$, so

$$\begin{aligned}\mu&=\sqrt{2}\{(\phi_1-\phi_s)^{\frac{1}{2}}+(\phi_2-\phi_s)^{\frac{1}{2}}\}\\ &\geqslant\sqrt{2}\,\delta^{\frac{1}{2}}\end{aligned} \tag{2.36}$$

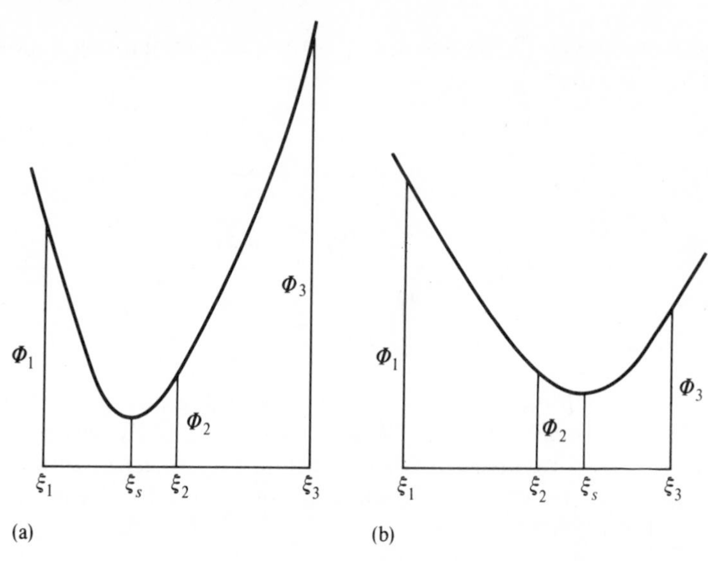

Fig. 2.8. The two cases in Powell's method.

since

$$\phi_1 = f_1 \geqslant f_2 \geqslant \cdots \geqslant f_{n+1} = \phi_2 > \phi_s.$$

So (2.33) does not hold, and no change should be made.

(b) Otherwise $\phi_1 > \phi_3$, so that $\boldsymbol{\xi}_s$ lies beyond $\boldsymbol{\xi}_2$ and then (2.35) becomes

$$\|\boldsymbol{\xi}_s - \boldsymbol{\xi}_1\| = \mu_1\sigma, \quad \mu_1 = \sqrt{2}(\phi_1 - \phi_s)^{\frac{1}{2}},$$
$$\|\boldsymbol{\xi}_s - \boldsymbol{\xi}_2\| = \mu_2\sigma, \quad \mu_2 = \sqrt{2}(\phi_2 - \phi_s)^{\frac{1}{2}},$$

and now (2.36) is replaced by

$$\mu = \sqrt{2}\{(\phi_1 - \phi_s)^{\frac{1}{2}} - (\phi_2 - \phi_s)^{\frac{1}{2}}\}. \tag{2.37}$$

Substituting from (2.34) for f_s gives μ in terms of ϕ_1, ϕ_2, ϕ_3 and condition (2.33) becomes, after rearrangement

$$(\phi_1 - 2\phi_2 + \phi_3)(\phi_1 - \phi_2 - \delta)^2 < \tfrac{1}{2}\delta(f_1 - f_3)^2. \tag{2.38}$$

This gives a complete set of conditions covering all cases.

The algorithm can now be summarised.

(1) From the current point, $\mathbf{x}_1$, with current value f_1, minimise successively along the current set of directions, $\mathbf{s}_1, \mathbf{s}_2, \ldots, \mathbf{s}_n$, finding points $\mathbf{x}_2, \mathbf{x}_3, \ldots, \mathbf{x}_{n+1}$, and values $f_2, f_3, \ldots, f_{n+1}$. Find the

maximum change along any direction, that is find

$$\max_j (f_j - f_{j+1}) = \delta,$$

corresponding to the direction $\mathbf{s}_{j_0}$.

(2) Write $\phi_1 = f(\mathbf{x}_1)$, $\phi_2 = f(\mathbf{x}_{n+1})$, $\phi_3 = f(2\mathbf{x}_{n+1} - \mathbf{x}_1)$. If $\phi_3 < \phi_2$, take the new current point as $2\mathbf{x}_{n+1} - \mathbf{x}_1$, otherwise take it as $\mathbf{x}_{n+1}$.

(3) Now test whether a direction $\mathbf{s}$ should be changed in the next cycle.

(a) If

$$\phi_1 - 2\phi_2 + \phi_3 < 0$$

or if

$$\phi_1 - 2\phi_2 + \phi_3 > 0, \quad \phi_1 > \phi_3,$$

$$\text{and} \quad (\phi_1 - 2\phi_2 + \phi_3)(\phi_1 - \phi_2 - \delta)^2 < \tfrac{1}{2}\delta(\phi_1 - \phi_5)^2,$$

then drop the direction $\mathbf{s}_{j_0}$, that is the direction giving the greatest change in f. Replace it with the direction $\mathbf{s}$, given by

$$\mathbf{s} = \mathbf{x}_{n+1} - \mathbf{x}_1$$

and minimise in the next cycle along directions $\mathbf{s}_1, \mathbf{s}_2, \ldots, \mathbf{s}_{j_0-1}, \mathbf{s}_{j_0+1}, \ldots, \mathbf{s}_n, \mathbf{s}$.

(b) If

$$\phi_1 - 2\phi_2 + \phi_3 > 0 \quad \text{and} \quad \phi_3 > \phi_1$$

or if

$$\phi_1 - 2\phi_2 + \phi_3 > 0, \quad \phi_1 > \phi_3,$$

$$\text{and} \quad (\phi_1 - 2\phi_2 + \varphi_3)(\varphi_1 - \varphi_2 - \delta)^2 > \tfrac{1}{2}\delta(\phi_1 - \phi_3)^2,$$

then make no change in the minimising directions.

(4) Return to (1) with the new current point (from (2)), and the directions $\mathbf{s}$ (from (3)).

The procedure terminates in a finite number of steps when f is quadratic since the directions used will ultimately be conjugate ones.

Example 2.6 As an example, consider minimising the function in Example 2.5

$$f(x_1, x_2) = (x_2 - x_1^2)^2 + (1 - x_1)^2$$

starting at the origin, and using initially the directions of the axes.
Cycle 1 $\mathbf{x}_1^{\mathrm{T}} = (0, 0)$, $f_1 = 1$, $\mathbf{s}_1^{\mathrm{T}} = (1, 0)$, $\mathbf{s}_2^{\mathrm{T}} = (0, 1)$; minimise $f(\mathbf{x}_1 + \alpha\mathbf{s}_1)$ to get $\mathbf{x}_2^{\mathrm{T}} = (0.5898, 0)$, $f_2 = 0.2893$; minimise $f(\mathbf{x}_2 + \alpha\mathbf{s}_2)$

to get $\mathbf{x}_3^T = (0.5898, 0.3479)$, $f_3 = 0.1683$. Now $(f_1 - f_2) = 0.7107$, $(f_2 - f_3) = 0.1210$, so $\delta = 0.7107$. Also $\boldsymbol{\xi}_1 = \mathbf{x}_1$, $\boldsymbol{\xi}_2 = \mathbf{x}_3$, $\boldsymbol{\xi}_3 = 2\mathbf{x}_3 - \mathbf{x}_1$ so $\boldsymbol{\xi}_3 = (1.1796, 0.6958)$ and $\phi_1 = f_1$, $\phi_2 = f_3$, $\phi_3 = f(\boldsymbol{\xi}_3) = 1.0001$. Since $\phi_3 > \phi_1$, repeat the cycle using the same directions and starting at $\mathbf{x}_3$.
Cycle 2 $\mathbf{x}_1^T = (0.5898, 0.3479)$, $f_1 = 0.1683$, $\mathbf{s}_1^T = (1, 0)$, $\mathbf{s}_2^T = (0, 1)$; minimising along $\mathbf{s}_1$ and $\mathbf{s}_2$ successively gives

$$\mathbf{x}_2^T = (0.7300, 0.3479), \quad f_2 = 0.1414,$$
$$\mathbf{x}_3^T = (0.7300, 0.5329), \quad f_3 = 0.0729.$$

Now $\delta = 0.0685$, and $\boldsymbol{\xi}_1 = \mathbf{x}_1$, $\boldsymbol{\xi}_2 = \mathbf{x}_3$, $\boldsymbol{\xi}_3^T = (0.8702, 0.7179)$, and so $\phi_1 = f_1$, $\phi_2 = f_3$, $\phi_3 = 0.0199 < \phi_1$. Considering condition (2), we have

$$\phi_1 - 2\phi_2 + \phi_3 = 0.0424,$$
$$(\phi_1 - \phi_2 - \delta)^2 = 0.0007, \tfrac{1}{2}\delta(\phi_1 - \phi_3)^2 = 0.0008,$$

so condition (2) holds and in the next cycle we use $\boldsymbol{\xi}_2 - \boldsymbol{\xi}_1$ as a direction and drop $\mathbf{s}_1$. Since $\phi_3 < \phi_2$ start the next cycle at $\boldsymbol{\xi}_3$.
Cycle 3

$$\mathbf{x}_1 = \begin{bmatrix} 0.8702 \\ 0.7179 \end{bmatrix}, \quad f_1 = 0.0199, \quad \mathbf{s}_1 = \begin{bmatrix} 0 \\ 1 \end{bmatrix}, \quad \mathbf{s}_2 = \begin{bmatrix} 0.1402 \\ 0.1850 \end{bmatrix}$$

and so on.

2.4 Newton and quasi Newton methods

2.4.1 Newton's method

The usual Newton–Raphson method of solving the equation $F(x) = 0$ in a scalar variable x is to carry out the iteration

$$x_{r+1} = x_r - F(x_r)/F'(x_r)$$

based on linearising the function about x_r. The corresponding method for finding a minimum of $f(\mathbf{x})$ by solving the vector equation $\mathbf{g}(\mathbf{x}) = 0$ is

$$\mathbf{x}_{r+1} = \mathbf{x}_r - \{\mathbf{G}(\mathbf{x}_r)\}^{-1}\mathbf{g}(\mathbf{x}_r), \tag{2.39}$$

where $\mathbf{g}(\mathbf{x})$ is the gradient and $\mathbf{G}(\mathbf{x})$ the Hessian matrix

$$G_{ij}(\mathbf{x}) = \frac{\partial^2 f}{\partial x_i \, \partial x_j}. \tag{2.40}$$

Near a minimum $\mathbf{G}$ is positive definite. For f quadratic, (2.8), $\mathbf{G}$ is constant and (2.39) produces the minimum, $\mathbf{x}^* = \mathbf{G}^{-1}\mathbf{b}$, in one step, from any initial point, with no requirement to calculate the step length. However, for general f the method requires the evaluation of $\frac{1}{2}n(n+1)$ second derivatives at each step and also a matrix inversion or solution of n linear equations, so that it is not a practicable procedure. The practical methods work by producing a succession of positive definite matrices $\mathbf{H}_r$ which behave like $\mathbf{G}^{-1}$ near the minimum, and at the r-th step they use search directions $\mathbf{p}_r = -\mathbf{H}_r^{\mathrm{T}}\mathbf{g}_r$.

Rate of convergence of descent processes The change in the value of the function at successive points may be examined in the simplest case, namely

$$f(\mathbf{x}) = \tfrac{1}{2}\mathbf{x}^{\mathrm{T}}\mathbf{G}\mathbf{x}.$$

Using steepest descents, the step from $\mathbf{x}_r$ is in the direction $-\mathbf{g}_r = -\mathbf{G}\mathbf{x}_r$ and

$$\mathbf{x}_{r+1} = \mathbf{x}_r - \alpha\mathbf{G}\mathbf{x}_r,$$

where α is such that $\mathbf{g}_{r+1}^{\mathrm{T}}\mathbf{g}_r = 0$. Solving for α and substituting gives, for this case, as in (2.14)

$$f(\mathbf{x}_{r+1})/f(\mathbf{x}_r) = 1 - (\mathbf{g}_r^{\mathrm{T}}\mathbf{g}_r)^2/(\mathbf{g}_r^{\mathrm{T}}\mathbf{G}\mathbf{g}_r)(\mathbf{g}_r^{\mathrm{T}}\mathbf{G}^{-1}\mathbf{g}_r).$$

By the Kantorovitch inequality (see Note 2.1), we can bound this in the case when $\mathbf{G}$ is positive definite with distinct eigenvalues; if λ_1 and λ_n are the least and greatest of these eigenvalues,

$$0 \leqslant f(\mathbf{x}_{r+1})/f(\mathbf{x}_r) \leqslant (\lambda_n/\lambda_1 - 1)^2/(\lambda_n/\lambda_1 + 1)^2.$$

If this ratio tends to a limit as $r \to \infty$, then the values of f will ultimately decrease by a common factor at each step, and the value of this factor can be bounded, with a smaller bound when the eigenvalues of $\mathbf{G}$ are more nearly equal.

The situation is quite different for a process like the Newton one. If we apply this to the quadratic function the minimum is reached in one step; so consider the function

$$f(\mathbf{x}) = \tfrac{1}{2}\mathbf{x}^{\mathrm{T}}\mathbf{G}\mathbf{x} + O(\|\mathbf{x}\|^3);$$

then

$$\mathbf{g}(\mathbf{x}_r) = \mathbf{G}\mathbf{x}_r + O(\|\mathbf{x}\|^2)$$

and

$$\mathbf{x}_{r+1} = \mathbf{x}_r - \mathbf{G}^{-1}\mathbf{g}_r = O(\|\mathbf{x}\|^2).$$

It follows that

$$f(\mathbf{x}_{r+1}) \text{ depends now on } (f(\mathbf{x}_r))^2$$

and so the Newton process gives a value now decreasing quadratically. For the simple function above, $f^* = 0$; the results given hold for the deviations $f(\mathbf{x}_r) - f^*$ with general functions f which satisfy certain conditions. There is an extensive theory establishing these, but it will not be considered further here.

2.4.2 Davidon–Fletcher–Powell method

We now describe what has so far proved to be the most successful of this class of algorithms. The method is defined by the following steps, $\mathbf{H}$ being an $n \times n$ matrix.

(1) $\mathbf{H}_1 = \mathbf{I}$
(2) $\mathbf{p}_r = -\mathbf{H}_r\mathbf{g}_r$
(3) $\mathbf{x}_{r+1}$ is the minimum of f along the line $\mathbf{x}_r + \alpha\mathbf{p}_r$
(4) $\mathbf{s}_r = \mathbf{x}_{r+1} - \mathbf{x}_r$, $\mathbf{y}_r = \mathbf{g}_{r+1} - \mathbf{g}_r$, and
$\mathbf{H}_{r+1} = \mathbf{H}_r + (1/\mathbf{s}_r^{\mathrm{T}}\mathbf{y}_r)\mathbf{s}_r\mathbf{s}_r^{\mathrm{T}} - (1/\mathbf{y}_r^{\mathrm{T}}\mathbf{H}_r\mathbf{y}_r)\mathbf{H}_r\mathbf{y}_r)\mathbf{y}_r^{\mathrm{T}}\mathbf{H}_r$.

The process then repeats from step (2).

This process has the following properties:

(1) All the $\mathbf{H}$'s are symmetric, a result proved immediately by induction since $\mathbf{H}_1$ is symmetric and $\mathbf{H}_{r+1} - \mathbf{H}_r$ is symmetric if $\mathbf{H}_r$ is.

(2) All the $\mathbf{H}$'s are positive definite. We prove this also by induction, noting that it is true for $\mathbf{H}_1$, and so assume it true for $\mathbf{H}_r$; then for any vector $\mathbf{z}$

$$\begin{aligned}\mathbf{z}^{\mathrm{T}}\mathbf{H}_{r+1}\mathbf{z} &= \mathbf{z}^{\mathrm{T}}\mathbf{H}_r\mathbf{z} + (1/\mathbf{s}_r^{\mathrm{T}}\mathbf{y}_r)\mathbf{z}^{\mathrm{T}}\mathbf{s}_r\mathbf{s}_r^{\mathrm{T}}\mathbf{z} \\ &\quad - (1/\mathbf{y}_r^{\mathrm{T}}\mathbf{H}_r\mathbf{y}_r)\mathbf{z}^{\mathrm{T}}\mathbf{H}_r\mathbf{y}_r\mathbf{y}_r^{\mathrm{T}}\mathbf{H}_r\mathbf{z} \\ &= (1/\mathbf{y}_r^{\mathrm{T}}\mathbf{H}_r\mathbf{y}_r)\{(\mathbf{z}^{\mathrm{T}}\mathbf{H}_r\mathbf{z})(\mathbf{y}_r^{\mathrm{T}}\mathbf{H}_r\mathbf{y}_r) - (\mathbf{z}^{\mathrm{T}}\mathbf{H}_r\mathbf{y}_r)^2\} \\ &\quad + (1/\mathbf{s}_r^{\mathrm{T}}\mathbf{y}_r)(\mathbf{z}^{\mathrm{T}}\mathbf{s}_r)^2. \end{aligned} \tag{2.41}$$

Since $\mathbf{H}_r$ is positive definite, $\mathbf{g}_r^{\mathrm{T}}\mathbf{H}_r\mathbf{g}_r = -\mathbf{p}_r^{\mathrm{T}}\mathbf{g}_r > 0$, so that $\mathbf{p}_r$ has a component in the direction of $-\mathbf{g}_r$; that is, f is reduced by a step $\mathbf{s}_r$ in the direction of $\mathbf{p}_r$. But $\mathbf{s}_r^{\mathrm{T}}\mathbf{y}_r = \mathbf{s}_r^{\mathrm{T}}(\mathbf{g}_{r+1} - \mathbf{g}_r) = -\mathbf{s}_r^{\mathrm{T}}\mathbf{g}_r$, since $\mathbf{s}_r^{\mathrm{T}}\mathbf{g}_{r+1} = 0$ and so $\mathbf{s}_r^{\mathrm{T}}\mathbf{y}_r > 0$. The second term in (2.41) is then positive.

In the first term of (2.41), since $\mathbf{H}_r$ is symmetric positive definite it can be written as

$$\mathbf{H}_r = \mathbf{L}_r \mathbf{L}_r^{\mathrm{T}},$$

where $\mathbf{L}$ is a real lower triangular matrix; substituting $\mathbf{z}^{\mathrm{T}}\mathbf{L}_r = \boldsymbol{\zeta}^{\mathrm{T}}$, $\mathbf{y}_r^{\mathrm{T}}\mathbf{L}_r = \boldsymbol{\eta}^{\mathrm{T}}$, this gives

$$(\mathbf{z}^{\mathrm{T}}\mathbf{H}_r\mathbf{z})(\mathbf{y}_r^{\mathrm{T}}\mathbf{H}_r\mathbf{y}_r) - (\mathbf{z}^{\mathrm{T}}\mathbf{H}_r\mathbf{y}_r)^2 = (\boldsymbol{\zeta}^{\mathrm{T}}\boldsymbol{\zeta})(\boldsymbol{\eta}^{\mathrm{T}}\boldsymbol{\eta}) - (\boldsymbol{\zeta}^{\mathrm{T}}\boldsymbol{\eta})^2,$$

which is positive by the Cauchy–Schwarz inequality.

Hence both terms on the right-hand side of (2.41) are positive (since $\mathbf{y}_r^{\mathrm{T}}\mathbf{H}_r\mathbf{y}_r > 0$) and so $\mathbf{z}^{\mathrm{T}}\mathbf{H}_{r+1}\mathbf{z} > 0$ for any $\mathbf{z}$, i.e. $\mathbf{H}_{r+1}$ is positive definite if $\mathbf{H}_r$ is.

This property guarantees that proceeding along each of the $\mathbf{p}$'s decreases f whatever the behaviour of f in the vicinity.

(3) The $\mathbf{p}$'s are conjugate directions when f is quadratic, and also

$$\mathbf{H}_r\mathbf{y}_K = \mathbf{s}_K \quad \text{for } K < r. \tag{2.42}$$

These two results can be proved together, again by induction. Initially

$$\mathbf{H}_2\mathbf{y}_1 = \mathbf{H}_1\mathbf{y}_1 + (1/\mathbf{s}_1^{\mathrm{T}}\mathbf{y}_1)\mathbf{s}_1\mathbf{s}_1^{\mathrm{T}}\mathbf{y}_1 - (1/\mathbf{y}_1^{\mathrm{T}}\mathbf{H}\mathbf{y}_1)\mathbf{H}_1\mathbf{y}_1\mathbf{y}_1^{\mathrm{T}}\mathbf{H}_1\mathbf{y}_1, \tag{2.43}$$

and $\mathbf{H}_1 = \mathbf{I}$, so $\mathbf{H}_2\mathbf{y}_1 = \mathbf{y}_1 + \mathbf{s}_1 - \mathbf{y}_1 = \mathbf{s}_1$. Also when f is quadratic with constant Hessian $\mathbf{G}$,

$$\mathbf{y}_1 = \mathbf{g}_2 - \mathbf{g}_1 = \mathbf{G}(\mathbf{x}_2 - \mathbf{x}_1) = \mathbf{G}\mathbf{s}_1,$$

and so

$$\mathbf{H}_2\mathbf{G}\mathbf{s}_1 = \mathbf{s}_1 \quad \text{or equivalently} \quad \mathbf{H}_2\mathbf{G}\mathbf{p}_1 = \mathbf{p}_1. \tag{2.44}$$

Now

$$\begin{aligned}
\mathbf{p}_2^{\mathrm{T}}\mathbf{G}\mathbf{p}_1 &= -\mathbf{g}_2^{\mathrm{T}}\mathbf{H}_2\mathbf{G}\mathbf{p}_1 && (\text{since } \mathbf{p}_2 = -\mathbf{H}_2\mathbf{g}_2)\\
&= -\mathbf{g}_2^{\mathrm{T}}\mathbf{p}_1 && (\text{from (2.44)})\\
&= 0 && (\text{since } \mathbf{x}_2 \text{ minimises } f \text{ along } \mathbf{p}_1).
\end{aligned}$$

Hence both the statements are true for $r = 2$. Now assume them true up to r. Then

$$\begin{aligned}
\mathbf{H}_{r+1}\mathbf{y}_r &= \mathbf{H}_r\mathbf{y}_r + (1/\mathbf{s}_r^{\mathrm{T}}\mathbf{y}_r)\mathbf{s}_r\mathbf{s}_r^{\mathrm{T}}\mathbf{y}_r - (1/\mathbf{y}_r^{\mathrm{T}}\mathbf{H}_r\mathbf{y}_r)\mathbf{H}_r\mathbf{y}_r\mathbf{y}_r^{\mathrm{T}}\mathbf{H}_r\mathbf{y}_r\\
&= \mathbf{H}_r\mathbf{y}_r + \mathbf{s}_r - \mathbf{H}_r\mathbf{y}_r = \mathbf{s}_r.
\end{aligned} \tag{2.45}$$

And

$$\begin{aligned}\mathbf{p}_{r+1}^{\mathrm{T}}\mathbf{G}\mathbf{p}_r &= -\mathbf{g}_{r+1}^{\mathrm{T}}\mathbf{H}_{r+1}\mathbf{G}\mathbf{p}_r \\ &= -\mathbf{g}_{r+1}^{\mathrm{T}}\mathbf{H}_{r+1}\mathbf{y}_r \\ &= -\mathbf{g}_{r+1}\mathbf{s}_r \quad \text{(from (2.45))} \\ &= 0 \end{aligned} \tag{2.46}$$

And for $K<r$,

$$\begin{aligned}\mathbf{H}_{r+1}\mathbf{y}_K &= \mathbf{H}_r\mathbf{y}_K + (1/\mathbf{s}_r^{\mathrm{T}}\mathbf{y}_r)\mathbf{s}_r\mathbf{s}_r^{\mathrm{T}}\mathbf{y}_K - (1/\mathbf{y}_r^{\mathrm{T}}\mathbf{H}_r\mathbf{y}_r)\mathbf{H}_r\mathbf{y}_r\mathbf{y}_r^{\mathrm{T}}\mathbf{H}_r\mathbf{y}_K \\ &= \mathbf{s}_K + (1/\mathbf{s}_r^{\mathrm{T}}\mathbf{y}_r)\mathbf{s}_r\mathbf{s}_r^{\mathrm{T}}\mathbf{G}\mathbf{s}_K - (1/\mathbf{y}_r^{\mathrm{T}}\mathbf{H}_r\mathbf{y}_r)\mathbf{H}_r\mathbf{y}_r\mathbf{s}_r^{\mathrm{T}}\mathbf{G}\mathbf{s}_K\end{aligned}$$

since $\mathbf{H}_r\mathbf{y}_K = \mathbf{s}_K$ by the induction assumption, and $\mathbf{y}_r^{\mathrm{T}} = \mathbf{s}_r^{\mathrm{T}}\mathbf{G}$, so

$$\mathbf{H}_{r+1}\mathbf{y}_K = \mathbf{s}_K, \tag{2.47}$$

since $\mathbf{s}_r^{\mathrm{T}}\mathbf{G}\mathbf{s}_K = 0$ by the conjugacy assumption.
Also, for $K<r$,

$$\begin{aligned}\mathbf{p}_{r+1}^{\mathrm{T}}\mathbf{G}\mathbf{p}_K &= -\mathbf{g}_{r+1}^{\mathrm{T}}\mathbf{H}_{r+1}\mathbf{G}\mathbf{p}_K \\ &= -\mathbf{g}_{r+1}^{\mathrm{T}}\mathbf{H}_{r+1}\mathbf{y}_K/\alpha_K \quad (\text{since } \mathbf{G}\mathbf{p}_K = \mathbf{y}_K/\alpha_K) \\ &= -\mathbf{g}_{r+1}^{\mathrm{T}}\mathbf{s}_K/\alpha_K \quad \text{(from (2.47))} \\ &= 0,\end{aligned} \tag{2.48}$$

since $\mathbf{p}_1, \ldots, \mathbf{p}_r$ are assumed conjugate and it is a general property (2.22) that $\mathbf{g}_{r+1}$ is then orthogonal to all the previous $\mathbf{p}$'s. The results (2.45)–(2.48) then prove the original statements.

(4) As a direct consequence of (2.42) and the expression for $\mathbf{y}_K$, we have

$$\mathbf{H}_r\mathbf{G}\mathbf{s}_K = \mathbf{s}_K \quad \text{for } K<r. \tag{2.49}$$

This means that $\mathbf{H}_r$ is the inverse of $\mathbf{G}$ on the subspace spanned by $\mathbf{p}_1, \ldots, \mathbf{p}_{r-1}$ and so, since $\mathbf{p}_1, \ldots, \mathbf{p}_n$ are independent (being conjugate)

$$\mathbf{H}_{n+1} = \mathbf{G}^{-1}. \tag{2.50}$$

Thus the direction of the steps changes from that of steepest descent $-\mathbf{g}_1$ at the start to the Newton direction $-\mathbf{G}^{-1}\mathbf{g}$ at the end. For a quadratic f the algorithm terminates with $\mathbf{g}=0$ in at most n steps. For a general f it is usual to restart every n-th step with the current gradient.

Example 2.7 To minimise the same function as in Examples 2.5, 2.6,

$$f(x_1, x_2) = (x_2 - x_1^2)^2 + (1-x_1)^2.$$

Cycle 1 $\mathbf{x}_1^T = (0, 0)$, $\mathbf{g}_1^T = (-2, 0)$, $\mathbf{p}_1^T = (2, 0)$ since $\mathbf{H}_1 = \mathbf{I}$. Minimising $f(\mathbf{x}_1 + \alpha\mathbf{p}_1)$ gives $\mathbf{x}_2^T = (0.5898, 0)$, $f_2 = 0.2893$, $\mathbf{g}_2^T = (0, -0.6957)$. Hence

$$\mathbf{s}_1^T = (\mathbf{x}_2 - \mathbf{x}_1)^T = (0.5898, 0), \quad \mathbf{y}_1^T = (\mathbf{g}_2 - \mathbf{g}_1)^T = (2, -0.6957),$$

and

$$\mathbf{s}_1^T\mathbf{y}_1 = 1.1796, \quad \mathbf{s}_1\mathbf{s}_1^T = \begin{bmatrix} 0.3479 & 0 \\ 0 & 0 \end{bmatrix},$$

and

$$\mathbf{s}_1\mathbf{s}_1^T/\mathbf{s}_1^T\mathbf{y}_1 = \begin{bmatrix} 0.2949 & 0 \\ 0 & 0 \end{bmatrix},$$

and

$$\mathbf{y}_1^T\mathbf{H}_1\mathbf{y}_1 = 4.4840, \quad \mathbf{H}_1\mathbf{y}_1\mathbf{y}_1^T\mathbf{H}_1 = \begin{bmatrix} 4 & -1.3914 \\ -1.3914 & 0.4840 \end{bmatrix},$$

so that

$$-\mathbf{H}_1\mathbf{y}_1\mathbf{y}_1^T\mathbf{H}_1/\mathbf{y}_1^T\mathbf{H}_1\mathbf{y}_1 = \begin{bmatrix} -0.8921 & 0.3103 \\ 0.3103 & -0.1079 \end{bmatrix}$$

and

$$\mathbf{H}_2 = \begin{bmatrix} 0.4028 & 0.3103 \\ 0.3103 & 0.8921 \end{bmatrix}.$$

This completes the first cycle.
Cycle 2 $\mathbf{x}_2^T = (0.5898, 0)$, $\mathbf{g}_2^T = (0, -0.6957)$,
$\mathbf{p}_2^T = -\mathbf{H}_2\mathbf{g}_2 = (0.2159, 0.6206)$ and the process continues by finding $\mathbf{x}_3$ by minimising along $\mathbf{p}_2$.

2.4.3 General considerations on matrix updating methods: the Huang family

In the DFP method the increment $\mathbf{H}_{r+1} - \mathbf{H}_r$ is a matrix of rank $-$ 2. A systematic way of developing a family of algorithms with similar properties was given by Huang (1970) and is sketched here.

Suppose we specify the following requirements:

(a) line searches only, that is each step requiring minimisation with respect to one variable;

(b) quadratic termination;
(c) use of function and gradient values only;
(d) search direction $\mathbf{p}_{r+1}$ dependent only on values of function and gradient at that point and the previous point, that is $\mathbf{H}_{r+1}$ dependent only on $\mathbf{x}_{r+1}$, $\mathbf{x}_r$, $\mathbf{g}_{r+1}$, $\mathbf{g}_r$.

The conditions (a) and (b) require the $\mathbf{p}$'s to be conjugate directions, that is

$$\mathbf{p}_r^{\mathrm{T}}\mathbf{G}\mathbf{p}_s = 0 \quad \text{for } r \neq s.$$

If we write, as in Section 2.4.2, $\mathbf{s}_r = \mathbf{x}_{r+1} - \mathbf{x}_r$, $\mathbf{y}_r = \mathbf{g}_{r+1} - \mathbf{g}_r$, and note that $\mathbf{y}_r = \mathbf{G}\mathbf{s}_r$, $\mathbf{s}_r = \alpha_r\mathbf{p}_r$, and $\mathbf{g}_r^{\mathrm{T}}\mathbf{p}_s = 0$ for $s < r$, then clearly

$$\mathbf{s}_r^{\mathrm{T}}\mathbf{G}\mathbf{s}_s = 0 \quad \text{for } r < s, \tag{2.51}$$

$$\mathbf{g}_r^{\mathrm{T}}\mathbf{s}_s = 0 \quad \text{for } s < r, \tag{2.52}$$

$$\mathbf{s}_r^{\mathrm{T}}\mathbf{y}_s = 0 \quad \text{for } r \neq s. \tag{2.53}$$

Now assume that the $\mathbf{p}$'s are generated from

$$\mathbf{p}_r = -\mathbf{H}_r^{\mathrm{T}}\mathbf{g}_r. \tag{2.54}$$

We try to produce a form for $\mathbf{H}_r$ which satisfies the requirements without explicitly involving $\mathbf{G}$. From (2.54) and conjugacy

$$\mathbf{g}_r^{\mathrm{T}}\mathbf{H}_r\mathbf{G}\mathbf{s}_s = 0 \quad \text{for } r \neq s \tag{2.55}$$

The simplest way to satisfy (2.55) and (2.52) is to write

$$\mathbf{H}_r\mathbf{G}\mathbf{s}_s = \rho\mathbf{s}_s \quad \text{for } s < r \tag{2.56}$$

with ρ a scalar constant. Note that this then implies

$$\mathbf{H}_{n+1}\mathbf{G}\mathbf{p}_s = \mathbf{p}_s \quad \text{for } s \leqslant n \tag{2.57}$$

and so $\mathbf{H}_{n+1} = \rho\mathbf{G}^{-1}$ since $\mathbf{p}_1, \mathbf{p}_2, \ldots, \mathbf{p}_n$ form a basis. From (2.56) and substituting $\mathbf{y}_s = \mathbf{G}\mathbf{s}_s$, we have

$$\mathbf{H}_r\mathbf{y}_s = \rho\mathbf{s}_s \quad \text{for } s < r \tag{2.58}$$

and so

$$(\mathbf{H}_{r+1} - \mathbf{H}_r)\mathbf{y}_s = 0 \quad \text{for } s < r, \tag{2.59}$$

$$(\mathbf{H}_{r+1} - \mathbf{H}_r)\mathbf{y}_r = \rho\mathbf{s}_r - \mathbf{H}_r\mathbf{y}_r. \tag{2.60}$$

Also from (2.58), $\mathbf{y}_r^{\mathrm{T}}\mathbf{H}_r\mathbf{y}_s = \rho\mathbf{y}_r^{\mathrm{T}}\mathbf{s}_s = 0$ for $s < r$ from (2.53). From this result and from (2.59), (2.60) it follows that $\mathbf{H}_{r+1} - \mathbf{H}_r$ can be expressed in terms of $\mathbf{s}_r$ and of $\mathbf{H}_r\mathbf{y}_r$; so consider the expression

$$\mathbf{H}_{r+1} - \mathbf{H}_r = \rho(1/\boldsymbol{\eta}_r^{\mathrm{T}}\mathbf{y}_r)\mathbf{s}_r\boldsymbol{\eta}_r^{\mathrm{T}} - (1/\boldsymbol{\zeta}_r^{\mathrm{T}}\mathbf{y})\mathbf{H}_r\mathbf{y}_r\boldsymbol{\zeta}_r^{\mathrm{T}}, \tag{2.61}$$

where

$$\boldsymbol{\eta}_r = C_1\mathbf{s}_r + C_2\mathbf{H}_r^{\mathrm{T}}\mathbf{y}_r, \quad \boldsymbol{\zeta}_r = K_1\mathbf{s}_r + K_2\mathbf{H}_r^{\mathrm{T}}\mathbf{y}_r \tag{2.62}$$

and C_1, C_2, K_1, K_2 are all arbitrary scalar constants. It can be seen that this satisfies (2.59) and also (2.60) for all values of C, K, ρ and it may be regarded as defining a family of rank -2 updating formulae. With any initial matrix $\mathbf{H}_1$ and point $\mathbf{x}_1$, (2.61) defines, with (2.54), a set of conjugate directions $\mathbf{p}$ and a set of matrices $\mathbf{H}$ which end with $\mathbf{H}_{n+1} = \rho\mathbf{G}^{-1}$. The DFP formula is found by putting $C_1 = 1$, $C_2 = 0$, $K_1 = 0$, $K_2 = 1$, $\rho = 1$ and $\mathbf{H}_1 = \mathbf{I}$.

This family has the very attractive property that for a given initial point $\mathbf{x}_1$ and initial matrix $\mathbf{H}_1$ all the algorithms generate the same search directions, that is the directions of the $\mathbf{p}$'s are independent of the values chosen for C, K and ρ. For a general proof of this see Dixon (1972); we sketch in the next section a proof for f quadratic.

2.4.4 Common directions for the Huang family

We prove first that (2.61) and (2.54) give

$$\mathbf{p}_r = \beta_r\mathbf{q}_r, \tag{2.63}$$

where

$$\beta_r = 1 - (K_2\mathbf{y}_{r-1}^{\mathrm{T}}\mathbf{H}_{r-1}^{\mathrm{T}}\mathbf{g}_r)/(\boldsymbol{\zeta}_{r-1}^{\mathrm{T}}\mathbf{y}_{r-1}) \tag{2.64}$$

and

$$\mathbf{q}_r = -(\mathbf{I} - \mathbf{s}_{r-1}\mathbf{y}_{r-1}^{\mathrm{T}}/\mathbf{s}_{r-1}^{\mathrm{T}}\mathbf{y}_{r-1})\mathbf{H}_{r-1}^{\mathrm{T}}\mathbf{g}_r. \tag{2.65}$$

Direct substitution from (2.54), (2.61) and (2.62) gives

$$\mathbf{p}_r = -\mathbf{H}_{r-1}^{\mathrm{T}}\mathbf{g}_r + \frac{(K_1\mathbf{s}_{r-1} - \mathbf{y}_{r-1}^{\mathrm{T}}\mathbf{H}_{r-1}^{\mathrm{T}}\mathbf{g}_r + K_2\mathbf{H}_{r-1}^{\mathrm{T}}\mathbf{y}_{r-1}\mathbf{y}_{r-1}^{\mathrm{T}}\mathbf{H}_{r-1}^{\mathrm{T}}\mathbf{g}_r)}{\boldsymbol{\zeta}_{r-1}^{\mathrm{T}}\mathbf{y}_{r-1}}$$

since $\mathbf{s}_{r-1}^{\mathrm{T}}\mathbf{g}_r = 0$. Substituting for $\mathbf{y}_{r-1}$ and then for $\mathbf{H}_{r-1}^{\mathrm{T}}\mathbf{g}_{r-1}$ gives

$$\mathbf{p}_r = -\beta_r\mathbf{H}_{r-1}^{\mathrm{T}}\mathbf{g}_r + \mathbf{y}_{r-1}^{\mathrm{T}}\mathbf{H}_{r-1}^{\mathrm{T}}\mathbf{g}_r(K_1\mathbf{s}_{r-1} + K_2\mathbf{p}_{r-1})/\boldsymbol{\zeta}_{r-1}^{\mathrm{T}}\mathbf{y}_{r-1}$$

and, from (2.64), substituting for $\boldsymbol{\zeta}_{r-1}$ and noting that $\mathbf{s}_{r-1} = \alpha_{r-1}\mathbf{p}_{r-1}$, $\mathbf{p}_r = \beta_r\mathbf{q}_r$ as required.

We now prove a general expression for $\mathbf{q}_r$:

$$\mathbf{q}_r = -\left(\mathbf{I} - \sum_{i=1}^{r-1} \mathbf{s}_i\mathbf{y}_i^{\mathrm{T}}/\mathbf{s}_i^{\mathrm{T}}\mathbf{y}_i\right)\mathbf{H}_1^{\mathrm{T}}\mathbf{g}_r. \tag{2.66}$$

First show that

$$\mathbf{H}_{r-1}^{\mathrm{T}}\mathbf{g}_r = (I - \mathbf{s}_{r-2}\mathbf{y}_{r-2}^{\mathrm{T}}/\mathbf{s}_{r-2}^{\mathrm{T}}\mathbf{y}_{r-2})\mathbf{H}_{r-2}^{\mathrm{T}}\mathbf{g}_r. \tag{2.67}$$

This follows by directly substituting as above using $\mathbf{s}_{r-2}^{\mathrm{T}}\mathbf{g}_r = 0$, $\mathbf{y}_{r-1}^{\mathrm{T}}\mathbf{s}_{r-2} = 0$, and

$$\mathbf{H}_{r-2}^{\mathrm{T}}\mathbf{g}_{r-1} = \frac{\mathbf{s}_{r-1}\mathbf{y}_{r-1}^{\mathrm{T}}\mathbf{H}_{r-2}^{\mathrm{T}}\mathbf{g}_{r-1}}{\mathbf{s}_{r-1}^{\mathrm{T}}\mathbf{y}_{r-1}} + \frac{\mathbf{s}_{r-2}\mathbf{y}_{r-2}^{\mathrm{T}}\mathbf{H}_{r-2}^{\mathrm{T}}\mathbf{g}_{r-1}}{\mathbf{s}_{r-2}^{\mathrm{T}}\mathbf{y}_{r-2}},$$

which follows from (2.63), (2.64), (2.65) with $r-1$ in place of r. Then (2.65), (2.67) give

$$\begin{aligned}\mathbf{q}_r &= (\mathbf{I} - \mathbf{s}_{r-1}\mathbf{y}_{r-1}^{\mathrm{T}}/\mathbf{s}_{r-1}^{\mathrm{T}}\mathbf{y}_{r-1})(\mathbf{I} - \mathbf{s}_{r-2}\mathbf{y}_{r-2}^{\mathrm{T}}/\mathbf{s}_{r-2}^{\mathrm{T}}\mathbf{y}_{r-2})\mathbf{H}_{r-2}^{\mathrm{T}}\mathbf{g}_r \\ &= -(\mathbf{I} - \mathbf{s}_{r-1}\mathbf{y}_{r-1}^{\mathrm{T}}/\mathbf{s}_{r-1}^{\mathrm{T}}\mathbf{y}_{r-1} - \mathbf{s}_{r-2}\mathbf{y}_{r-2}^{\mathrm{T}}/\mathbf{s}_{r-2}^{\mathrm{T}}\mathbf{y}_{r-2})\mathbf{H}_{r-2}^{\mathrm{T}}\mathbf{g}_r,\end{aligned}$$

since $\mathbf{y}_{r-1}^{\mathrm{T}}\mathbf{s}_{r-2} = 0$. The process can be continued to produce (2.66) as required.

As a consequence of this, all algorithms of the Huang family produce the same sequence of points when started with a common $\mathbf{x}_1$ and $\mathbf{H}_1$. This follows because the first minimisation is along $-\mathbf{H}_1^{\mathrm{T}}\mathbf{g}_1$, and gives $\mathbf{x}_2$ and $\mathbf{g}_2$. Then, as above, $\mathbf{p}_2$ is along $\mathbf{q}_2$, and $\mathbf{q}_2$ depends only on $\mathbf{y}_1 = \mathbf{g}_2 - \mathbf{g}_1$, $\mathbf{s}_1 = \mathbf{x}_2 - \mathbf{x}_1$ and $\mathbf{H}_1$, $\mathbf{g}_2$ and so is the same for all members of the family. By the same argument, the same sequence of $\mathbf{x}_3, \mathbf{x}_4, \ldots$ is generated whatever the values of ρ, C_1, C_2, K_1 and K_2.

2.4.5 The conjugate-gradient algorithm revisited

From the results (2.63), (2.66) we can show that the conjugate-gradient algorithm also reproduces the same search directions and so the same points. Putting $\mathbf{H}_1 = \mathbf{I}$ so that $\mathbf{q}_1 = -\mathbf{g}_1$, (2.66) becomes

$$\begin{aligned}\mathbf{q}_{r+1} &= -\left(\mathbf{I} - \sum_{i=1}^{r} \mathbf{s}_i\mathbf{y}_i^{\mathrm{T}}/\mathbf{s}_i^{\mathrm{T}}\mathbf{y}_i\right)\mathbf{g}_{r+1} \\ &= -\mathbf{g}_{r+1} + \sum_{i=1}^{r} \mathbf{s}_i\mathbf{y}_i^{\mathrm{T}}\mathbf{g}_{r+1}/\mathbf{s}_i^{\mathrm{T}}\mathbf{y}_i \\ &= -\mathbf{g}_{r+1} + \mathbf{s}_r\mathbf{g}_{r+1}^{\mathrm{T}}\mathbf{g}_{r+1}/\mathbf{s}_r^{\mathrm{T}}\mathbf{y}_r,\end{aligned}$$

since $\mathbf{g}_p^{\mathrm{T}}\mathbf{g}_{r+1} = 0$, $p < r+1$.

Now $\mathbf{s}_r$ is along $\mathbf{q}_r$, and

$$\begin{aligned}\mathbf{q}_r^{\mathrm{T}}\mathbf{y}_r &= -\mathbf{g}_r^{\mathrm{T}}\left(\mathbf{y}_r - \sum_{i=1}^{r-1} \mathbf{y}_i\mathbf{s}_i^{\mathrm{T}}\mathbf{y}_r/\mathbf{s}_i^{\mathrm{T}}\mathbf{y}_i\right) \\ &= \mathbf{g}_r^{\mathrm{T}}\mathbf{g}_r \quad \text{since} \quad \mathbf{s}_i^{\mathrm{T}}\mathbf{y}_r = 0\end{aligned}$$

and (2.66) becomes

$$\mathbf{q}_{r+1} = -\mathbf{g}_{r+1} + (\mathbf{g}_{r+1}^{\mathrm{T}}\mathbf{g}_{r+1}/\mathbf{g}_r^{\mathrm{T}}\mathbf{g}_r)\mathbf{q}_r, \tag{2.68}$$

which is simply the conjugate-gradient algorithm of Section 2.3.3.

2.4.6 Other methods

Besides the rank -2 methods derived in Section 2.4.2 there are rank -1 methods (see, for example, Problem 12) and combinations of these. One idea is to drop the requirement of exact line search while retaining descent properties and so produce a more robust algorithm. The practical difficulties which arise are usually associated with $\mathbf{H}_r$ becoming singular when the subsequent points become confined to a subspace not, in general, including the minimum. It is on these grounds that algorithms which are theoretically precisely equivalent can differ in their performance.

One adaptation of the Newton method may be mentioned here since it is particularly suitable for a special class of problem, namely those where $f(\mathbf{x})$ is a sum of squared terms. These problems arise when fitting by least squares. The Newton method (for general f) is

$$\mathbf{x}_{r+1} = \mathbf{x}_r - \{\mathbf{G}(\mathbf{x}_r)\}^{-1}\mathbf{g}(\mathbf{x}_r).$$

If now

$$f(\mathbf{x}) = \tfrac{1}{2}\sum_{i=1}^{m}\{y_i - h_i(\mathbf{x})\}^2 \tag{2.69}$$

then a comparable method can be produced by linearising $h_i(\mathbf{x})$ about the current point $\mathbf{x}_r$, that is by writing

$$\begin{aligned} f(\mathbf{x}_r + \boldsymbol{\delta}) &= \tfrac{1}{2}\sum_{i=1}^{m}\{y_i - h_i(\mathbf{x}_r) - \nabla h_i(\mathbf{x}_r)^{\mathrm{T}}\boldsymbol{\delta}\}^2 \\ &= \tfrac{1}{2}(\mathbf{Y}_r - \mathbf{P}_r\boldsymbol{\delta})^{\mathrm{T}}(\mathbf{Y}_r - \mathbf{P}_r\boldsymbol{\delta}) \end{aligned} \tag{2.70}$$

where $\mathbf{P}_r(\mathbf{x})$ is the $m \times n$ matrix whose rows are $\nabla h_i(\mathbf{x}_r)^{\mathrm{T}}$ and $\mathbf{Y}_r(\mathbf{x})$ is the m-vector $\mathbf{y} - \mathbf{h}(\mathbf{x}_r)$. Minimising f with respect to $\boldsymbol{\delta}$ then gives

$$\nabla f = \mathbf{g}(\mathbf{x}_r + \boldsymbol{\delta}) = -\mathbf{P}_r^{\mathrm{T}}\mathbf{Y}_r + \mathbf{P}_r^{\mathrm{T}}\mathbf{P}_r\boldsymbol{\lambda} = 0$$

and so

$$\boldsymbol{\delta}_r = (\mathbf{P}_r^{\mathrm{T}}\mathbf{P}_r)^{-1}\mathbf{P}_r^{\mathrm{T}}\mathbf{Y}_r \tag{2.71}$$

and $\mathbf{x}_{r+1} = \mathbf{x}_r + \boldsymbol{\delta}_r$. (2.71) is sometimes called the generalised least-squares method. Unfortunately it often fails to converge. An improvement (Levenberg (1944), Marquardt (1963)) can be produced by limiting the size of the correction $\boldsymbol{\delta}_r$. If we impose the condition

$$\|\boldsymbol{\delta}_r\|^2 = \delta_0^2 \tag{2.72}$$

and minimise $f(\mathbf{x}_r + \boldsymbol{\delta})$ with respect to $\boldsymbol{\delta}$ subject to this condition, then we require

$$\mathbf{P}_r^{\mathrm{T}}(\mathbf{Y}_r - \mathbf{P}_r\boldsymbol{\delta}_r) - \lambda\boldsymbol{\delta}_r = 0,$$

λ being the Lagrange multiplier attached to condition (2.72); that is,

$$(\mathbf{P}_r^{\mathrm{T}}\mathbf{P}_r + \lambda\mathbf{I})\boldsymbol{\delta} = \mathbf{P}_r^{\mathrm{T}}\mathbf{Y}_r. \tag{2.73}$$

Another way of deriving this same condition is to note that the approximate Hessian $\mathbf{P}_r^{\mathrm{T}}\mathbf{P}_r$ should be "sufficiently" positive definite for (2.71) to behave satisfactorily, and that the correction in (2.73), for positive λ, is a way of ensuring this. In practice λ is altered during the process; when $\lambda = 0$ the Newton step is produced, while a large λ corresponds to a short step which in the limit is in the direction $\mathbf{P}_r^{\mathrm{T}}\mathbf{Y}_r$, that is $-\mathbf{g}_r$, the steepest-descent step. Implementations of this method differ in the way the parameter is defined.

A similar idea could be used with Newton's method applied to a general function f, but then the Hessian would be required; the advantage of the least-squares linearising is that it provides second-order behaviour while only requiring gradients.

2.5 Summary of methods

The methods described in this section for minimising a general function f are steepest descents (Section 2.3.1), conjugate gradients (Section 2.5.3), Powell's method (Section 2.3.4), Newton's method (Section 2.4.1), Davidon–Fletcher–Powell (Section 2.4.2), and the Levenberg–Marquardt algorithm for least-squares minimisation (Section 2.4.6).

Note

2.1 Kantorovitch inequality

We state and prove the simplest form only. If $\mathbf{A}$ is a symmetric positive definite matrix, it has real positive eigenvalues; suppose these to be distinct;

$$\lambda_1 < \lambda_2 < \cdots < \lambda_n.$$

Then

$$1 \leqslant (\mathbf{x}^T\mathbf{A}\mathbf{x})(\mathbf{x}^T\mathbf{A}^{-1}\mathbf{x})/(\mathbf{x}^T\mathbf{x})^2 \leqslant (\lambda_1 + \lambda_n)^2/4\lambda_1\lambda_n.$$

Proof Since there exists an orthogonal transformation

$$\mathbf{x} = \mathbf{P}\mathbf{y}$$

which converts $\mathbf{A}$ to $\mathbf{P}^T\mathbf{A}\mathbf{P}$, the diagonal matrix

$$\mathbf{\Lambda} = \begin{bmatrix} \lambda_1 & & 0 \\ & \lambda_2 & \\ 0 & & \ddots & \\ & & & \lambda_n \end{bmatrix},$$

we can without loss of generality take $\mathbf{A}$ to be of this form initially; and then $\mathbf{A}^{-1}$ is also diagonal, with diagonal elements $1/\lambda_i$, and

$$\mathbf{x}^T\mathbf{A}\mathbf{x} = \sum_{i=1}^{n} \lambda_i \mathbf{x}_i^2, \quad \mathbf{x}^T\mathbf{A}^{-1}\mathbf{x} = \sum_{i=1}^{n} x_i^2/\lambda_i.$$

So now consider the quantity

$$\left(\sum_{i=1}^{n} \lambda_i x_i^2\right)\left(\sum_{i=1}^{n} x_i^2/\lambda_i\right) \Big/ \left(\sum_{i=1}^{n} x_i^2\right)^2,$$

which can be written

$$\left(\sum_{i=1}^{n} w_i\lambda_i\right)\left(\sum_{i=1}^{n} w_i/\lambda_i\right)$$

with $w_i \geqslant 0, \sum_{i=1}^{n} w_i = 1$. Clearly this is $\geqslant 1$. with equality when any one of the weights w_i is unity; and if $w_1 = w_n = \frac{1}{2}$, it takes the value $(\lambda_1 + \lambda_n)^2/4\lambda_1\lambda_n$. Then by considering varying these weights, it can be verified that this is the maximum value.

Problems

(1) Apply the methods of Fibonacci and golden section to find the minimum of $f(x) = x^3 - 24x + 3$ within the interval $1 < x < 3.5$ correct to ± 0.05.

(2) Prove that for a fixed large value of N, the golden section method produces an interval of uncertainty about 17% greater than the Fibonacci method.

(3) Verify that the cubic

$$p_3(x) = (1+2x)(x-1)^2 f_0 + (3-2x)x^2 f_1 + (x-1)^2 x g_0 + x^2(x-1)g_1$$
$$= (1/6)P_3 x^3 + (1/2)P_2 x^2 + P_1 x + P_0$$

takes the values $p_3(0) = f_0$, $p_3(1) = f_1$, $p_3'(0) = g_0$, $p_3'(1) = g_1$. Show that the conditions (i) $g_0 < 0$, $g_1 > 0$ or (ii) $g_0 > 0$, $g_1 > 0$, $f_1 < f_0$ or (iii) $g_0 < 0$, $g_1 < 0$, $f_1 > f_0$ each guarantees that $p_3(x)$ has a minimum at a point x_m such that $0 < x_m < 1$. Show that such a minimum could exist also with $g_0 < 0$, $g_1 < 0$, $f_1 < f_0$ but that the extra condition $g_0 > P_2^2/(2P_3)$ is then necessary. Obtain the cubic $p_3(x)$ which fits the function and derivative at $x = 0$, $x = 1$ for the function $f(x) = -(3+4x)/(1+x^2)$. What is the minimum of this $p_3(x)$? What is the minimum of $f(x)$?

(4) Evaluate the first three points in the sequence produced by using steepest descents to minimise

$$f(\mathbf{x}) = 9x_1^2 - 8x_1x_2 + 3x_2^2$$

starting at $\mathbf{x}_1$, $x_1 = 1$, $x_2 = 1$.

Convert f by the transformation $\mathbf{x} = \mathbf{P}\mathbf{y}$, where P is an orthogonal matrix, to the form

$$g(\mathbf{y}) = \lambda_1 y_1^2 + \lambda_2 y_2^2.$$

Why does the sequence produced by minimising g by steepest descents starting at $\mathbf{y}_1 = \mathbf{P}^{\mathrm{T}}\mathbf{x}_1$ reproduce the points $\mathbf{P}^{\mathrm{T}}\mathbf{x}_2$, $\mathbf{P}^{\mathrm{T}}\mathbf{x}_3, \ldots$? Convert g by the substitution $y_1 = Y_1/\sqrt{\lambda_1}$, $y_2 = T_2/\sqrt{\lambda_2}$, to the form

$$G(\mathbf{Y}) = Y_1^2 + Y_2^2.$$

Verify that the minimum of G is reached in one step for any initial point.

(5) For given $f(\mathbf{x}) = \frac{1}{2}\mathbf{x}^{\mathrm{T}}\mathbf{Q}\mathbf{x} + \mathbf{b}^{\mathrm{T}}\mathbf{x}$, with positive definite matrix $\mathbf{Q}$ and minimum point $\mathbf{x}^*$, we define

$$E(\mathbf{x}) = \tfrac{1}{2}(\mathbf{x} - \mathbf{x}^*)^{\mathrm{T}}\mathbf{Q}(\mathbf{x} - \mathbf{x}^*).$$

Prove that, if one steepest descent step is from $\mathbf{x}_K$ to $\mathbf{x}_{K+1}$, then

$$\{E(\mathbf{x}_K)-E(\mathbf{x}_{K+1})\}/E(\mathbf{x}_K)=r_K=(\mathbf{g}_K^T\mathbf{g}_K)^2/\{(\mathbf{g}_K^T\mathbf{Q}\mathbf{g}_K)(\mathbf{g}_K^T\mathbf{Q}^{-1}\mathbf{g}_K)\}.$$

The value of r_K is a measure of the relative rate of convergence of the process at $\mathbf{x}_K$. Take $\mathbf{Q}$ as a diagonal matrix $\mathbf{\Lambda}$ (why is this possible without loss of generality?) and then $r_K=r_K(\mathbf{\Lambda})$. Show that

(i) $r=1$ if and only if $\mathbf{Q}$ is the identity matrix

(ii) if Λ_1 has diagonal elements $\lambda_1<\lambda_2<\lambda_3<\cdots<\lambda_n$, and $\mathbf{\Lambda}_2$ has the same set of diagonal elements with λ_1 replaced by λ'_1, $\lambda_1<\lambda'_1<\lambda_2$, and λ_n replaced by $\lambda'_n\lambda_{n+1}<\lambda'_n<\lambda_n$, then

$$r(\mathbf{\Lambda}_1)<r(\mathbf{\Lambda}_2)$$

(6) For the quadratic objective function

$$f(\mathbf{x})=2x_1^2+x_2^2/2+5x_3^2/2+x_1x_2-2x_1x_3-8x_1-3x_2+7x_3$$

the method of conjugate gradients starting at $\mathbf{x}_1$, where $\mathbf{x}_1^T=(0,0,0)$, proceeds via $\mathbf{x}_2^T=(1.248, 0.468, -1.092)$ and $\mathbf{x}_3^T=(1.491, 1.084, -0.726)$. Verify that $\mathbf{g}_3^T\mathbf{p}_2=\mathbf{g}_3^T\mathbf{p}_1=\mathbf{g}_2^T\mathbf{p}_1=0$. Complete the minimisation by finding $\mathbf{x}_4(=\mathbf{x}^*)$.

(7) Find the stationary points of the function

$$f(\mathbf{x})=x_1^3+x_1x_2+x_1^2x_2^2-3x_1$$

by solving the equations $g_1(\mathbf{x})=0$, $g_2(\mathbf{x})=0$, and decide their nature by evaluating the Hessian $\mathbf{G}(\mathbf{x})$ at each.

Try to minimise the function by using Newton's method starting at $\mathbf{x}_1$, where $\mathbf{x}_1^T=(1,1)$. Show that this converges but to a saddle point. Investigate the effect of replacing $\mathbf{G}^{-1}$ in the first step by $(\mathbf{G}+\lambda\mathbf{I})^{-1}$, taking $\lambda=2$.

(8) Find the sequence of points produced by Powell's method applied to minimising the function in Problem 6 with the same initial value $\mathbf{x}_1$.

(9) Prove that the eigenvectors of a real symmetric matrix $\mathbf{Q}$, corresponding to distinct eigenvalues, are conjugate directions with respect to $\mathbf{Q}$.

(10) In the DFP method it is possible to use

$$\mathbf{x}_{r+1}=\mathbf{x}_r+\alpha\mathbf{p}_r$$

with α a constant, instead of finding $\mathbf{x}_{r+1}$ by minimising along $\mathbf{p}_r$. Prove that all $\mathbf{H}_r$ are still positive definite.

(11) Let $\mathbf{Q}$ be the diagonal matrix

$$\mathbf{Q}=\begin{bmatrix}1 & 0 & 0\\ 0 & 2 & 0\\ 0 & 0 & 3\end{bmatrix}.$$

Consider minimising (a) $f_1(\mathbf{x})=\mathbf{x}^{\mathrm{T}}\mathbf{Q}\mathbf{x}$ and (b) $f_2(\mathbf{x})=(\mathbf{x}^{\mathrm{T}}\mathbf{Q}\mathbf{x})^2$, starting at $\mathbf{x}_1^{\mathrm{T}}=(1, 1, 1)$ and using conjugate directions. Take three steps with each function.

(12) The symmetric updating formula of rank -1 is

$$\mathbf{H}_{r+1}-\mathbf{H}_r=(\mathbf{s}_r-\mathbf{H}_r\mathbf{y}_r)(\mathbf{s}_r-\mathbf{H}_r\mathbf{y}_r)^{\mathrm{T}}/\{(\mathbf{s}_r-\mathbf{H}_r\mathbf{y}_r)^{\mathrm{T}}\mathbf{y}_r\}.$$

Show that

$$(\mathbf{H}_{r+1}-\mathbf{H}_r)\mathbf{y}_r=\mathbf{s}_r-\mathbf{H}_r\mathbf{y}_r$$

and that

$$(\mathbf{H}_{r+1}-\mathbf{H}_r)\mathbf{y}_s=0 \quad \text{for } s<r.$$

Hence show that, for $f(\mathbf{x})$ quadratic with constant Hessian $\mathbf{Q}$, the sequence of directions

$$\mathbf{p}_r=-\mathbf{H}_r\mathbf{g}_r$$

are $\mathbf{Q}$-conjugate and satisfy

$$\mathbf{H}_r\mathbf{Q}\mathbf{p}_s=\mathbf{p}_s \quad \text{for } s<r.$$

(13) In minimising a quadratic function of $\mathbf{x}\in R^n$,

$$f(\mathbf{x})=\tfrac{1}{2}\mathbf{x}^{\mathrm{T}}\mathbf{Q}\mathbf{x}+\mathbf{b}^{\mathrm{T}}\mathbf{x}$$

a set of conjugate directions $\mathbf{p}_r$ can be generated from the unit vectors along the axes by the formulae (2.17), with $\mathbf{u}_r=\mathbf{e}_r$, where $\mathbf{e}_r^{\mathrm{T}}=(0, 0, \ldots, 0, 1, 0, \ldots, 0)$ the 1 being the r-th component. Then $\mathbf{x}^*=\sum \alpha_r\mathbf{p}_r$ as in (2.19), with $\mathbf{x}_1$ taken at the origin. Show that

$$x_n^*=\alpha_n$$

and that the matrix with columns $\mathbf{Q}\mathbf{p}_1, \mathbf{Q}\mathbf{p}_2, \ldots, \mathbf{Q}\mathbf{p}_n$ is lower triangular. In fact this matrix, $\mathbf{L}$, is part of the triangular decomposition of $\mathbf{Q}$, so that

$$\mathbf{Q}=\mathbf{L}\mathbf{U}$$

where $\mathbf{U}$ is an upper triangular matrix with diagonal elements unity. Use these $\mathbf{p}_r$'s to minimise the quadratic function in Problem 6.

(14) The secant method for solving an equation in one variable, $g(x)=0$, is to evaluate g at two points x_1, x_2 and then, if $g(x_1)=g_1$, $g(x_2)=g_2$, and $x_2-x_1=s_1$, $g_2-g_1=y_1\neq 0$, to use the expression

$$x=x_2-s_1g_2/y_1$$

as an improvement on both x_1 and x_2. This can then be used as an iterative process. Generalise this to solving the equation

$$\mathbf{g}(\mathbf{x})=0$$

where $\mathbf{g}$, $\mathbf{x}$ are n-vectors, and show that the corresponding form is

$$\mathbf{x}=\mathbf{x}_{n+1}-\mathbf{S}\mathbf{Y}^{-1}\mathbf{g}_{n+1}$$

where $\mathbf{S}$, $\mathbf{Y}$ are $n\times n$ matrices whose columns are respectively $\mathbf{S}_r$, $\mathbf{y}_r$ where

$$\mathbf{S}_r=\mathbf{x}_{r+1}-\mathbf{x}_r, \quad \mathbf{y}_r=\mathbf{g}_{r+1}-\mathbf{g}_r, \quad r=1,2\cdots n$$

and $\mathbf{Y}$ is nonsingular.

Test this method by finding the minimum of the quadratic function in Problem 6 using as base points $\mathbf{x}_1^T=(0,0,0)$, $\mathbf{x}_2^T=(1,0,0)$, $\mathbf{x}_3^T=(0,1,0)$, and $\mathbf{x}_4^T=(0,0,1)$.

3

Linear programming

The simplest case of constrained optimisation is when the function and the constraints are all linear functions of $\mathbf{x}$. This is a model which is widely applicable to real life situations, at least as a first approximation, and for which the solution can always be obtained using an efficient general algorithm, the simplex method. This is developed in Section 3.1, and Section 3.2 is devoted to duality relations in the linear case.

3.1 Solution of LP problems

3.1.1 Statement of the problem

The general constrained problem of Chapter 1 becomes in the linear case

$$\text{minimise } f = \mathbf{g}^{\mathrm{T}}\mathbf{x} \text{ subject to } \mathbf{a}_i^{\mathrm{T}}\mathbf{x} \geqslant b_i, \quad i = 1, 2, \ldots, m \qquad (3.1)$$

where now $\mathbf{g}$, $\mathbf{a}_i$ are independent of $\mathbf{x}$. However it is often convenient to deal with the equivalent problem

$$\text{minimise } f = \mathbf{g}^{\mathrm{T}}\mathbf{x} \text{ subject to } \mathbf{A}\mathbf{x} = \mathbf{b}, \quad \mathbf{x} \geqslant 0. \qquad (3.2)$$

The non-negativity restriction on $\mathbf{x}$ frequently arises naturally from the conditions, and is built in to the solution algorithm. The two forms (3.1) and (3.2) are interchangeable though with different dimensions and different matrices $\mathbf{A}$, thus a constraint in (3.1)

$$\mathbf{a}_i^{\mathrm{T}}\mathbf{x} \geqslant b_i$$

can be replaced by two constraints, including one additional (slack) variable, in (3.2)

$$\mathbf{a}_i^{\mathrm{T}}\mathbf{x} - x_{n+i} = b_i, \quad x_{n+i} \geqslant 0$$

and similarly an unrestricted variable x_j in (3.1) may be replaced, for (3.2), by a difference of nonnegative variables

$$x_j = x_{j1} - x_{j2}, \quad x_{j1} \geqslant 0, \quad x_{j2} \geqslant 0.$$

Finally any equality constraint can be replaced by two inequality constraints so that

$$\mathbf{a}_i^{\mathrm{T}}\mathbf{x} = b_i$$

in (3.2) is equivalent to

$$\mathbf{a}_i^{\mathrm{T}}\mathbf{x} \geqslant b_i$$
$$-\mathbf{a}_i^{\mathrm{T}}\mathbf{x} \geqslant -b_i$$

in (3.1).

We shall use the form (3.2) as the standard LP problem in the rest of this section, with $\mathbf{x}$ an n-vector, $\mathbf{b}$ an m-vector, $m < n$, and, as before, shall call the solution set X, so that

$$X = \{\mathbf{x} \mid \mathbf{Ax} = \mathbf{b}, \mathbf{x} \geqslant 0\}.$$

3.1.2 Theory of linear problems

The general theory of Chapter 1 will now be specialised for this case. Two remarks may be made immediately.

(i) As in Section 1.2.5, X is a convex set, f is convex, and so a local minimum is also a global minimum and, if there is more than one minimum point, then f takes the same value everywhere on the convex hull of such points.
(ii) As in Section 1.2.1, minima can only lie on the boundary of X – since here the gradient $\mathbf{g}$ of f is constant, and so nonzero everywhere.

We note that the boundary is made up of hyperplanes. There are a finite number of corner points, i.e. points for which $n-m$ of the x_j are zero and such that the remaining m x_j, found by solving the equations $\mathbf{Ax} = \mathbf{b}$, turn out to be non-negative. We prove now that a linear objective function takes its minimum value only at these points, so that we need consider not even the whole boundary of X but only a finite set of points on this boundary. It is this fact which makes the linear problem so tractable. We need first a definition.

Definition An extreme point of a convex set K is a point which cannot be expressed as a linear combination of two distinct points of K. This is applicable to any convex set – in Fig. 3.1(a) P_1, P_2,

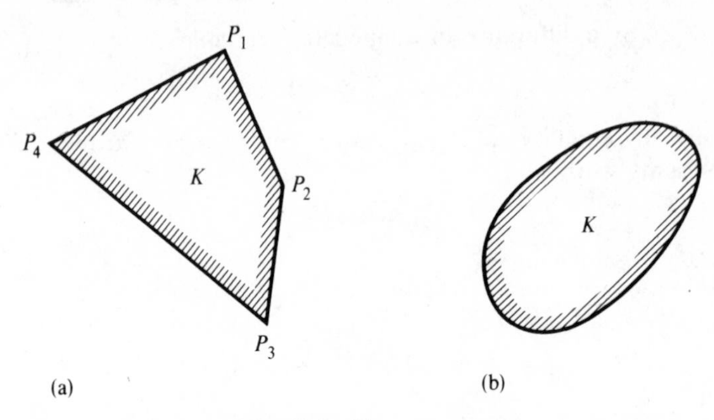

Fig. 3.1. Convex sets.

P_3 and P_4 are extreme points, in (b) every point on the boundary is extreme.

Now we can prove that the corner points of X are extreme, as would be expected. The point $\mathbf{x}$ is an extreme point of the solution set X iff $x_j > 0$ only for j such that the columns $\mathbf{a}^j$ of the matrix $\mathbf{A}$ are independent.

Proof Suppose $\mathbf{x}$ has $x_j > 0$, $j \in J$; since $\mathbf{x} \in X$,

$$\sum_{j \in J} x_j \mathbf{a}^j = \mathbf{b}. \tag{3.3}$$

If $\mathbf{x}$ is not extreme, then there exist $\mathbf{x}_1, \mathbf{x}_2 \in X$ and θ, $0 < \theta < 1$, such that

$$\mathbf{x} = \theta \mathbf{x}_1 + (1-\theta)\mathbf{x}_2. \tag{3.4}$$

It is clear from (3.4) that for $j \notin J$, since $x_j = 0$ and $\theta > 0$, $1-\theta > 0$, then $x_{1j} = 0$, $x_{2j} = 0$. And since $\mathbf{x}_1 \in X$, $\mathbf{x}_2 \in X$,

$$\sum_{j \in J} x_{1j} \mathbf{a}^j = \mathbf{b}, \quad \sum_{j \in J} x_{2j} \mathbf{a}^j = \mathbf{b}. \tag{3.5}$$

But then

$$\sum_{j \in J} (x_{1j} - x_{2j}) \mathbf{a}^j = 0,$$

and this is a contradiction since the $\mathbf{a}^j$ are independent. Hence $\mathbf{x}$ is extreme.

Conversely suppose $\mathbf{x}$ is extreme but that the $\mathbf{a}^j$ are dependent,

$j \in J$. Then there exist α_j not all zero such that

$$\sum_{j \in J} \alpha_j \mathbf{a}^j = 0. \tag{3.6}$$

But (3.6) together with (3.3) means that, for any μ

$$\sum_{j \in J} (x_j \pm \mu\alpha_j)\mathbf{a}^j = \mathbf{b}, \tag{3.7}$$

and since $x_j > 0$ it is certainly possible to choose μ so that $(x_j \pm \mu\alpha_j) > 0$, all $j \in J$. Hence (3.7) gives two feasible solutions

$$\mathbf{x}_1 = \mathbf{x} + \mu\boldsymbol{\alpha}, \quad \mathbf{x}_2 = \mathbf{x} - \mu\boldsymbol{\alpha}$$

and then $\mathbf{x} = \frac{1}{2}(\mathbf{x}_1 + \mathbf{x}_2)$, a contradiction since $\mathbf{x}$ is extreme. Hence the $\mathbf{a}^j$ are independent.

It is clear that an extreme point of X can have at most m nonzero components, since the rank of $\mathbf{A}$ is $\leqslant m$, and hence X has a finite number of extreme points. Also it follows directly from the definitions that a bounded convex set with a finite number of extreme points is the convex hull of these points. (Prove this.) It is of course possible that the solution set X is unbounded, or empty, so that these cases also will have to be considered. However for a bounded solution set we can now prove the main result.

A linear function defined on a bounded set K has a finite extreme value and takes that value at an extreme point of the set; if it has the same value at more than one extreme point, then it takes the same value on the convex hull of such points. (Note that this applies to a *general* convex set K.)

Proof Suppose that $f = \mathbf{g}^{\mathrm{T}}\mathbf{x}$ takes its minimum value f^* at $\mathbf{x}^*$, not an extreme point. Then there exist extreme points $\mathbf{x}_1, \mathbf{x}_2, \ldots, \mathbf{x}_p$ and $\mu_j > 0$ such that

$$\sum_{j=1}^{p} \mu_j = 1, \quad \mathbf{x}^* = \sum_{j=1}^{p} \mu_j \mathbf{x}_j. \tag{3.8}$$

But writing $f_j = \mathbf{g}^{\mathrm{T}}\mathbf{x}_j$, (3.8) means

$$f^* = \sum_{j=1}^{p} \mu_j f_j.$$

If the f_j are not all equal, then $f^* > \min_j f_j$, which is a contradiction: if the f_j are all equal, then their common value must be f^*, which confirms the second part of the statement.

It follows then that a linear function defined on a bounded solution set X takes its extreme value *either* at a unique extreme point, *or* at more than one extreme point, and then on the convex hull of such extremes; and that an extreme point is characterised by having its nonzero components associated with independent columns of $\mathbf{A}$. These points are called basic feasible solutions.

We now illustrate these statements on some simple problems.

Example 3.1

$$\begin{aligned} 3x_1+4x_2&\leqslant 12 \quad \text{i.e.} \quad & 3x_1+4x_2+x_3 \qquad &= 12\\ 2x_1+\ x_2&\leqslant 6 & 2x_1+\ x_2 \qquad +x_4 &= 6\\ x_1, x_2&\geqslant 0 & x_1, x_2, x_3, x_4 &\geqslant 0. \end{aligned}$$

The additional variables x_3, x_4 are known as slack variables. The set X is as shown in Fig. 3.2.

There are four extreme points,

$$\mathbf{x}_1^{\mathrm{T}}=(0,0,12,6), \qquad \mathbf{x}_2^{\mathrm{T}}=(3,0,3,0),$$
$$\mathbf{x}_3^{\mathrm{T}}=(12/5,6/5,0,0), \quad \mathbf{x}_4^{\mathrm{T}}=(0,3,0,3).$$

The set X is bounded, and is the convex hull of these four points. Each extreme point is associated with two independent columns of $\mathbf{A}$, thus $\mathbf{x}_1$ has as basis the columns of

$$\begin{bmatrix} 1 & 0\\ 0 & 1 \end{bmatrix},$$

$\mathbf{x}_2$ those of

$$\begin{bmatrix} 3 & 1\\ 2 & 0 \end{bmatrix}$$

and so on. Any linear objective will have its minimum value at one or more of these four points: thus $f=-x_1$ has a minimum $f^*=-3$ at $\mathbf{x}^*=\mathbf{x}_2$, and $f=-2x_1-x_2$ has a minimum value of -6 at all points of the form

$$\mathbf{x}=\theta\mathbf{x}_2+(1-\theta)\mathbf{x}_3, \quad 0\leqslant\theta\leqslant 1;$$

that is, all points on the line joining $\mathbf{x}_2$ and $\mathbf{x}_3$.

Example 3.2

$$\begin{aligned} 3x_1+4x_2&\leqslant 12 \quad \text{i.e.} \quad & 3x_1+4x_2+x_3 \qquad &= 12\\ 2x_1+\ x_2&\leqslant 3 & 2x_1+\ x_2 \qquad +x_4 &= 3\\ x_1, x_2&\geqslant 0 & x_1, x_2, x_3, x_4 &\geqslant 0. \end{aligned}$$

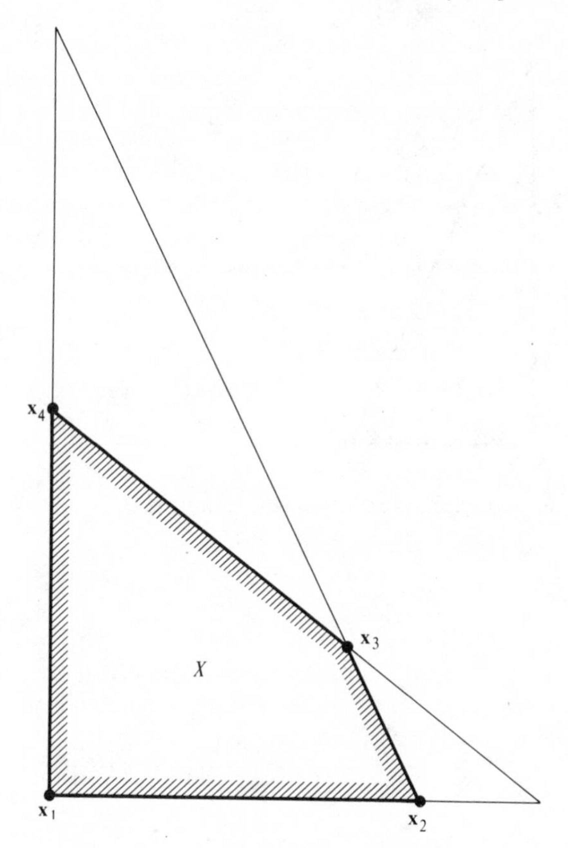

Fig. 3.2. Solution set, Example 3.1.

The solution set is shown in Fig. 3.3.

Here there are three extreme points

$$\mathbf{x}_1^{\mathrm{T}} = (0, 0, 12, 3), \quad \mathbf{x}_2^{\mathrm{T}} = (3/2, 0, 15/2, 0), \quad \mathbf{x}_3^{\mathrm{T}} = (0, 3, 0, 0).$$

The point $\mathbf{x}_3$ has only one nonzero component, multiplying the second column of $\mathbf{A}$; however we can associate any of the three other columns with this to form a basis for $\mathbf{x}_3$, that is a basis for $\mathbf{x}_3$ is given by the columns of any of the following:

$$\begin{bmatrix} 3 & 4 \\ 2 & 1 \end{bmatrix}, \begin{bmatrix} 4 & 1 \\ 1 & 0 \end{bmatrix}, \begin{bmatrix} 4 & 0 \\ 1 & 1 \end{bmatrix}.$$

Equivalently the point $\mathbf{x}_3$ results if we choose to make $x_3 = 0$,

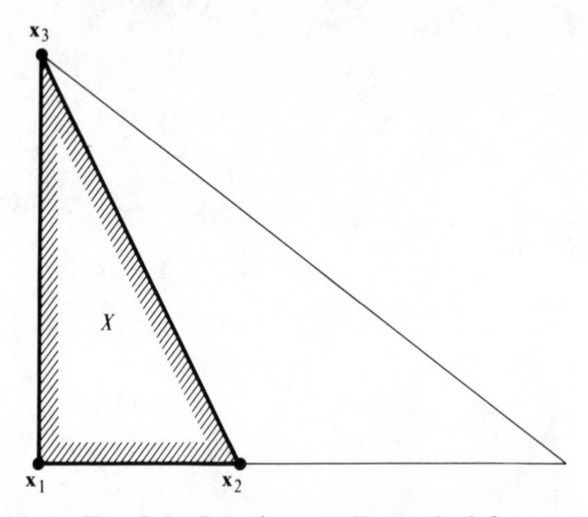

Fig. 3.3. Solution set, Example 3.2.

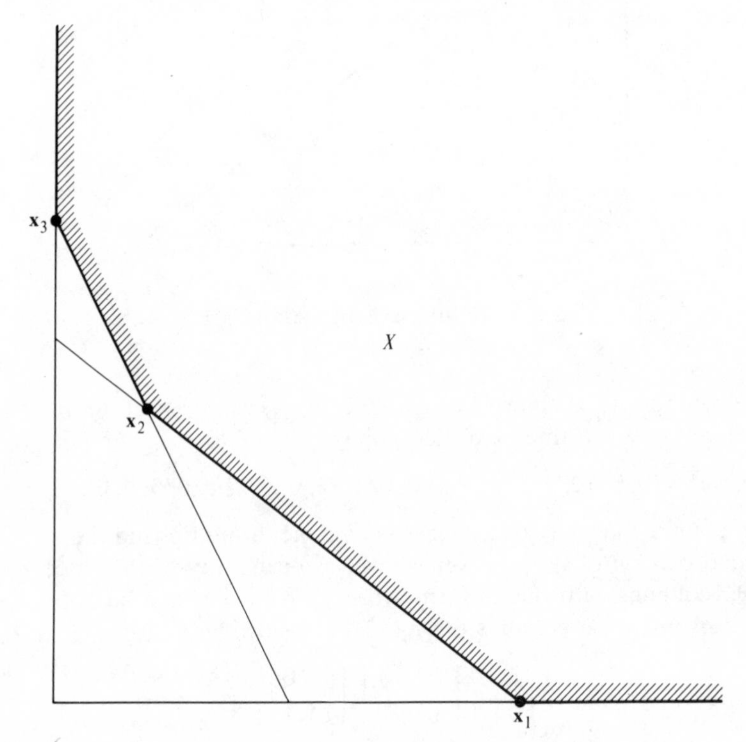

Fig. 3.4. Solution set, Example 3.3.

$x_4=0$ *or* $x_1=0$, $x_4=0$ *or* $x_1=0$, $x_3=0$. Such a point is called a degenerate basic feasible solution. Degeneracy does not alter the solution but it can affect the solution algorithm.

Example 3.3

$$\begin{array}{llll} 3x_1+4x_2\geqslant 12 & \text{i.e.} & 3x_1+4x_2-x_3 & =12 \\ 2x_1+\ x_2\geqslant\ 4 & & 2x_1+\ x_2 \qquad -x_4 & =\ 4 \\ x_1, x_2\geqslant\ 0 & & x_1, x_2, x_3, x_4 & \geqslant\ 0. \end{array}$$

This has the unbounded solution set X shown in Fig. 3.4.

There are three extreme points

$$\mathbf{x}_1^{\mathrm{T}}=(4,0,0,4), \quad \mathbf{x}_2^{\mathrm{T}}=(4/5,12/5,0,0), \quad \mathbf{x}_3^{\mathrm{T}}=(0,4,4;0).$$

It is now possible for a linear function to have a minimum or to be unbounded – for example $f=x_1+x_2$ has a minimum value of 16/5, at $\mathbf{x}_2$, but $f=-x_1-x_2$ is unbounded in X. (We shall show in the next section that if a minimum exists when X is unbounded, then it is at an extreme point, or a convex hull of extreme points.)

3.1.3 Simplex method

It is clear from the previous section that the only points we need to consider are extreme points, each characterised by a set of basic x's and the corresponding basis matrix made up of columns of $\mathbf{A}$. The simplex method generates a sequence of such points, ensuring at each step that the objective function is reduced (or at worst, not increased). The method changes only one column in the basis matrix at one time, so that one basic x_j becomes zero and another column joins the basis, its x_j changing from zero to positive. Since we can think of edges of the solution set X as defined by $n-m-1$ x's being zero, each such change corresponds to a step from one extreme point to an adjacent one along an edge of X. At a degenerate basic feasible solution a change of basis can be made without a change in position so that some special arrangement may be necessary to avoid getting into a loop.

The step from one point to the next corresponds to the step

$$\mathbf{x}_{r+1}=\mathbf{x}_r+\alpha_r\mathbf{p}_r$$

in nonlinear minimisation: the difference is that here there are only a finite number of directions $\mathbf{p}_r$, namely the edges of X

through $\mathbf{x}_r$, in which to go. We need to choose which of these is best, to proceed along it to the extreme point at its end, and to evaluate the new f and the new edge directions for the next step. The reason for choosing to proceed in this manner, one variable change at a time, is that the algorithm, the simplex method, is then a simple one.

Consider the situation at an extreme point $\mathbf{x}_r$ corresponding to a basis set $j \in J_r$, and a function value $f_r = \mathbf{g}^T\mathbf{x}_r$. It is convenient to transform the original problem

$$\mathbf{A}\mathbf{x} = \mathbf{b}, \quad -\mathbf{g}^T\mathbf{x} + f = 0 \tag{3.9}$$

into a form where each equation contains one and only one basic variable, and the objective function is related to the non-basic variables only. Note that this is equivalent to multiplying by an inverse matrix, though the usual method is designed to avoid having to do this explicitly. Assume that this has been done for $\mathbf{x}_r$, and that, for simplicity, the set J_r is the set $1, 2, \ldots, m$; then the resulting standard form is

$$\left.\begin{array}{llll}
x_1 & + s_{1,m+1}^{(r)} x_{m+1} + \cdots + s_{1,j}^{(r)} x_j + \cdots + s_{1,n}^{(r)} x_n & & = x_{r1} \\
\quad x_2 & + s_{2,m+1}^{(r)} x_{m+1} + \cdots + s_{2,j}^{(r)} x_j + \cdots + s_{2,n}^{(r)} x_n & & = x_{r2} \\
\cdot \quad \cdot \quad \cdot & \cdot \quad \cdot \quad \cdot \quad \cdot \quad \cdot \quad \cdot \quad \cdot \quad \cdot & & \cdot \\
\qquad x_m & + s_{m,m+1}^{(r)} x_{m+1} + \cdots + s_{m,j}^{(r)} x_j + \cdots + s_{m,n}^{(r)} x_n & & = x_{rm} \\
 & h_{m+1}^{(r)} x_{m+1} + \cdots + h_j^{(r)} x_j + \cdots + h_n^{(r)} x_n & + f & = f_r.
\end{array}\right\} \tag{3.10}$$

Note that this represents a feasible solution only if $x_{rj} \geqslant 0$, $j = 1, 2, \ldots, m$. A method of finding this, or of proving that no solution exists, will be described later: here it is assumed that this form has been found. The matrix of coefficients

$$\mathbf{T}_r = \left[\begin{array}{c|c} \mathbf{S}^{(r)} & \mathbf{x}_r \\ \hline \mathbf{h}^{(r)\mathrm{T}} & f_r \end{array}\right]$$

is frequently called a simplex tableau.

From (3.10) we now consider the effect of allowing a nonbasic x, that is an x_j, $m+1 \leqslant j \leqslant n$, to become positive. It is clear that

(i) if all $h_j^{(r)} \leqslant 0$, then such a change cannot decrease f. Hence f_r must then be the local minimum and so, by Section 3.1.2(i), the global minimum. If all $h_j^{(r)} < 0$, this is a unique global minimum.

(ii) if some $h_j^{(r)} > 0$, then making that x_j positive will decrease f. If more than one $h_j^{(r)} > 0$, it is usual to choose the largest, that is to move in the direction where f is decreasing at the greatest rate. So now consider making x_j a basic variable and dropping one of the present basic variables while keeping the other basic variables non-negative and the other nonbasic variables zero. This means considering only

$$\begin{array}{llll} x_1 & & + s_{1,j}^{(r)} x_j & = x_{r1} \\ & x_2 & + s_{2,j}^{(r)} x_j & = x_{r2} \\ \cdots & \cdots & \cdots & \cdots \\ & & x_m + s_{m,j}^{(r)} x_j & = x_{rm} \\ & & h_j^{(r)} x_j + f = f_r. & \end{array} \tag{3.11}$$

Suppose first that all $s_{ij}^{(r)} < 0$; then making x_j positive causes each basic x to increase and also, since $h_j^{(r)} > 0$, causes f to decrease, and each process can be continued without bound. Hence there is no finite solution. If however some $s_{ij}^{(r)} > 0$, then it is possible to reach an adjacent extreme point, $\mathbf{x}_{r+1}$, for which $x_{r+1,i} = 0$, $x_{r+1,j} > 0$. From (3.11), this means

$$x_{r+1,j} = x_{ri}/s_{ij}^{(r)} \tag{3.12}$$

and

$$\begin{aligned} x_{r+1,k} &= x_{rk} - s_{kj}^{(r)} x_{ri}/s_{ij}^{(r)}, \quad k \neq i, \quad 1 \leqslant k \leqslant m. \\ f_{r+1} &= f_r - h_j^{(r)} x_{ri}/s_{ij}^{(r)}. \end{aligned} \tag{3.13}$$

If more than one $s_{ij}^{(r)}$ is positive, then from (3.13) we need to choose i giving the smallest value of the ratio $x_{ri}/s_{ij}^{(r)}$, since this keeps the rest of the basic variables positive.

Having now chosen the new basis, with x_j replacing x_i, we need to produce a new tableau with the new basic variables, and the objective function, expressed in terms of the new nonbasic variables. We can do this directly. The (positive) element $s_{ij}^{(r)}$ is called the pivot element. Then

$$\begin{array}{lll} \text{(pivot row)} & x_i \;\; + s_{i,m+1}^{(r)} x_{m+1} + \cdots + s_{ij}^{(r)} x_j + \cdots & = x_{ri} \\ k\text{-th row} & x_k + s_{k,m+1}^{(r)} x_{m+1} + \cdots + s_{kj}^{(r)} x_j + \cdots & = x_{rk} \\ \text{objective row} & h_{m+1}^{(r)} x_{m+1} + \cdots + h_j^{(r)} x_j + \cdots + f & = f_r. \end{array}$$

Substituting for $x_{r+1,j}$, $x_{r+1,k}$ from (3.12), (3.13) and eliminating x_j between each other row and the pivot row produces

$$\begin{array}{ll} x_j + \{s_{i,m+1}^{(r)}/s_{ij}^{(r)}\} x_{m+1} + \cdots + \{1/s_{ij}^{(r)}\} \; x_i + \cdots & = x_{r+1,j} \\ x_k + \{s_{k,m+1}^{(r)} - s_{i,m+1}^{(r)} s_{kj}^{(r)}/s_{ij}^{(r)}\} x_{m+1} + \cdots - \{s_{kj}^{(r)}/s_{ij}^{(s)}\} x_i + \cdots & = x_{r+1,k} \\ \{h_{m+1}^{(r)} - s_{i,m+1}^{(r)} h_j^{(r)}/s_{ij}^{(r)}\} x_{m+1} + \cdots - \{h_j^{(r)}/s_{ij}^{(r)}\} x_i + \cdots + f & = f_{r+1}. \end{array} \tag{3.14}$$

The elimination process produces a common pattern for all these elements, and we shall call a term $\{s_{kp}s_{ij} - s_{ip}s_{kj}\}$ the "cross product" term for s_{kp} with pivot s_{ij}. Then the rule for obtaining the $(r+1)$-th tableau from the r-th is

(a) coefficients in the pivot row are divided by the pivot element;
(b) the pivot column is otherwise made up of zeros;
(c) any other element is replaced by its cross product divided by the pivot.

We can now summarise the steps of the simplex method.

(1) Obtain a basic feasible solution and produce the corresponding tableau. (Methods for doing this will be discussed later – sometimes it is obvious.)
(2) If all h_j are non-positive, the minimum is attained; if not, choose the largest positive h_j.
(3) If all $s_{ij} \leqslant 0$ for $h_j > 0$ the problem has an unbounded solution. If not, for fixed j choose among $s_{ij} > 0$ that i for which x_{ri}/s_{ij} is smallest. Then s_{ij} is the pivot, and the new tableau is found as above. Repeat from Step 2.

If at each step the function f is decreased, the process must terminate at the minimum, or indicate that this is unbounded. The only other possibility is that f is unchanged and the process cycles through a succession of degenerate solutions. This will be discussed in Section 3.1.6: first we give a simple example. (The f-column is omitted.)

Example 3.4 Minimise $f = -x_1 - x_2$ subject to

$$\begin{aligned} 3x_1 + 4x_2 + x_3 \quad\quad &= 12 \\ 2x_1 + x_2 \quad\quad + x_4 &= 6 \\ x_1, x_2, x_3, x_4 &\geqslant 0. \end{aligned}$$

(These are the conditions of Example 3.1.)

This problem is already in the simplex form with a first tableau $\mathbf{T}_1$ (Table 3.1) (corresponding to the point $\mathbf{x}_1$), and $f_1 = 0$.

Table 3.1

$\mathbf{T}_1$						
	x_3	3	4	1	0	12
	x_4	②	1	0	1	6
		1	1	0	0	0

Table 3.2

T_2						
x_3	0	(5/2)	1	−3/2		3
x_1	1	1/2	0	1/2		3
	0	1/2	0	−1/2		−3

Table 3.3

$T_3=T^*$						
x_2	0	1	2/5	−3/5		6/5
x_1	1	0	−1/5	4/5		12/5
	0	0	−1/5	−1/5		−18/5

Choose x_1 as the nonbasic variable to enter the basis, and, since $12/3>6/2$, take 2 as the pivot element, x_4 becoming zero. Then the new tableau is $\mathbf{T}_2$ (Table 3.2), which corresponds to the point $\mathbf{x}_2$ $(3, 0, 3, 0)$ and to $f_2=-3$. Repeating, we pivot on 5/2, making x_2 basic, x_3 nonbasic, and find $\mathbf{T}_3$ (Table 3.3). Since all h are now negative, the minimum is attained: it is at the point $\mathbf{x}_3$, $(12/5, 6/5, 0, 0)$ and has a function value $f_3=-18/5$.

3.1.4 Starting the simplex method

A starting solution is always obvious if the original problem is in the form $\mathbf{Ax}\leqslant\mathbf{b}$, with $\mathbf{b}\geqslant 0$, $\mathbf{x}\geqslant 0$. When this is not the case, a solution can be obtained by the device of introducing additional (surplus) non-negative variables and using the simplex method to minimise their sum; if the sum cannot be reduced to zero, then the original problem must have an empty solution set. The method is best explained by an example.

Example 3.5 Minimise $5x_1+6x_2+7x_3$ subject to

$$\begin{aligned} x_1+5x_2-3x_3&\geqslant 15\\ 5x_1-6x_2+10x_3&\leqslant 20\\ x_1+x_2+x_3&=5\\ x_1, x_2, x_3&\geqslant 0. \end{aligned}$$

The first step is to replace the inequalities so as to put the problem in the standard form (3.1). This gives, introducing slack variables x_4, x_5

$$\begin{aligned} x_1+5x_2-3x_3-x_4\qquad&=15\\ 5x_1-6x_2+10x_3\qquad+x_5&=20\\ x_1+x_2+x_3\qquad&=5\\ x_1, x_2, x_3, x_4, x_5&\geqslant 0. \end{aligned}$$

This is not in the form of a starting simplex tableau, so we introduce surplus variables and write the conditions

$$\begin{aligned} x_1+5x_2-\ 3x_3-x_4 \qquad +w_1 \qquad &= 15 \\ 5x_1-6x_2+10x_3 \qquad +x_5 \qquad\qquad &= 20 \quad \mathbf{x}, \mathbf{w} \geqslant 0 \\ x_1+\ x_2+\ \ x_3 \qquad\qquad\qquad +w_2 &= \ 5 \end{aligned}$$

together with the original objective, f, for which

$$-5x_1-6x_2-\ 7x_3 \qquad\qquad +f=\ 0$$

and a new objective made up of the surplus variables

$$-w_1-w_2+z=0.$$

If we now minimise z, and find a minimum of zero, this must mean w_1 and w_2 zero and hence give a solution of the original problem. The starting tableau will have w_1, x_5 and w_2 as basic variables, so we need to express z in terms of the other variables; adding the first and third equations to the equation for z produces

$$2x_1+6x_2-2x_3-x_4 \qquad +z=20.$$

The process is started with the tableau $\mathbf{T}_1$ (Table 3.4) and proceeds by the algorithm, carrying the f-row along with the rest, to the stages $\mathbf{T}_2$ (Table 3.5) and $\mathbf{T}_3$ (Table 3.6).

Table 3.4

$\mathbf{T}_1$									
w_1	1	(5)	−3	−1	0	1	0	15
x_5	5	−6	10	0	1	0	0	20
w_2	1	1	1	0	0	0	1	5
	−5	−6	−7	0	0	0	0	0
	2	6	−2	−1	0	0	0	20

Table 3.5

$\mathbf{T}_2$								
x_2	1/5	1	−3/5	−1/5	0	1/5	0	3
x_5	31/5	0	32/5	−6/5	1	6/5	0	38
w_2	4/5	0	(8/5)	1/5	0	−1/5	1	2
	−19/5	0	−53/5	−6/5	0	6/5	0	18
	4/5	0	8/5	1/5	0	−6/5	0	2

Table 3.6

$\mathbf{T}_3$							
x_2	1/2	1	0	−1/8	0		15/4
x_5	3	0	0	−2	1		30
x_3	(1/2)	0	1	1/8	0		5/4
	3/2	0	0	1/8	0		125/4

At this stage the first phase is completed. There is no need to calculate the z-row since zero is certainly the minimum, nor to calculate the w-columns. A basic solution $x_2=15/4$, $x_3=5/4$, $x_5=30$ has been found for the original problem, and the corresponding f value is 125/4. Pivoting to replace x_3 by x_1 produces $\mathbf{T}_4$ (Table 3.7), which is the optimum solution (5/2, 5/2, 0, 0, 75/2) with $f=55/2$.

Table 3.7

$\mathbf{T}_4=\mathbf{T}^*$						
x_2						5/2
x_5						75/2
x_1						5/2
	0	0	−3	−1/4	0	55/2

An alternative formulation is to write the constraints as above including slack and surplus variables but instead of minimising $z=w_1+w_2$ first, followed by f, to consider minimising

$$f_{\mathrm{A}}=f+M(w_1+w_2),$$

where M is a "very large" number. It is easy to show that this proceeds through exactly the same steps as the method above, since the minimisation starts by reducing the very large terms involving M to zero and then deals with the rest. If the minimum value involves M, then the original problem has no feasible solution.

3.1.5 Revised simplex method

In computer implementation it is convenient to carry through the operations of the simplex method in terms of the matrices involved, and a discussion of this will lead naturally to a treatment of duality in Section 3.2. We revert to (3.10) and write the tableau in matrix form as

$$\mathbf{T}_r:\left[\begin{array}{c|c|c|c}\mathbf{I}_m & \mathbf{S}^{(r)} & 0 & \mathbf{x}_{br}\\ \hline 0 & \mathbf{h}^{(r)\mathrm{T}} & 1 & f_r\end{array}\right], \tag{3.15}$$

where $\mathbf{S}^{(r)}$ is $m\times(n-m)$, and $\mathbf{x}_{br}$ is the column m-vector containing the m basic components (in general nonzero) of $\mathbf{x}_r$. To obtain the next tableau $\mathbf{T}_{r+1}$ from this one we need to multiply by a matrix inverse to

$$\mathbf{M}_r = \left[\begin{array}{ccccc} 1 & & s_{1j}^{(r)} & & 0 \\ & 1 & s_{2j}^{(r)} & & 0 \\ & & s_{ij}^{(r)} & & 0 \\ & & s_{mj}^{(r)} & 1 & 0 \\ \hline 0 & 0\cdots & h_j^{(r)} & \cdots 0 & 1 \end{array}\right] \tag{3.16}$$

You can confirm that this inversion has a simple form: $\{\mathbf{M}_r\}^{-1}$ is formed by replacing the pivot element $s_{ij}^{(r)}$ by $1/s_{ij}^{(r)}$, every other element in that column, $s_{kj}^{(r)}$, by $-s_{kj}^{(r)}/s_{ij}^{(r)}$, and $h_j^{(r)}$ by $-h_j^{(r)}/s_{ij}^{(r)}$. By considering the whole succession of these operations, the tableau at any stage is a product of the initial tableau matrix and the successive inverse matrices. The advantage is that the whole tableau need not be produced at each stage; inspection of the $h^{(r)}$ shows which variable should become basic, calculation of that column only gives the pivot, and the inverse matrix product $\prod_{k=1}^{r}\{\mathbf{M}_k\}^{-1}$ can be updated. As an illustration, consider again the problem already solved in Section 3.1.3.

Example 3.6 Solve the problem in Example 3.4 by the revised simplex method. The first tableau $\mathbf{T}_1$ is shown in Table 3.8. By

Table 3.8

$\mathbf{T}_1$						
x_3	3	4	1	0	12
x_4	②	1	0	1	6
	1	1	0	0	0.

the usual argument we find the pivot element, 2. Then the next tableau will arise by multiplying by the inverse of the matrix

$$\mathbf{M}_1 = \begin{bmatrix} 1 & 3 & 0 \\ 0 & 2 & 0 \\ 0 & 1 & 1 \end{bmatrix},$$

where the second is the pivot column, the pivot element being on the diagonal, and the third column represents the coefficient of f

which is always 1. As above

$$\{\mathbf{M}_1\}^{-1} = \begin{bmatrix} 1 & -3/2 & 0 \\ 0 & 1/2 & 0 \\ 0 & -1/2 & 1 \end{bmatrix}.$$

Now $\{\mathbf{M}_1\}^{-1}\mathbf{T}_1$ produces $\mathbf{T}_2$, but suppose we only scan the values of $\mathbf{h}$:

$$\mathbf{h}^{(2)\mathrm{T}} = [0 \quad -1/2 \quad 1] \begin{bmatrix} 4 & 0 \\ 1 & 1 \\ 1 & 0 \end{bmatrix} = [1/2 \quad -1/2].$$

This indicates that the minimum has not been reached, and that x_2 should enter the basis. Now scan $\mathbf{s}^2$ and $\mathbf{x}_{b2}$:

$$\mathbf{s}^2 = \begin{bmatrix} 1 & -3/2 \\ 0 & 1/2 \end{bmatrix} \begin{bmatrix} 4 \\ 1 \end{bmatrix} = \begin{bmatrix} 5/2 \\ 1/2 \end{bmatrix},$$

$$\mathbf{x}_{b2} = \begin{bmatrix} 1 & -3/2 \\ 0 & 1/2 \end{bmatrix} \begin{bmatrix} 12 \\ 6 \end{bmatrix} = \begin{bmatrix} 3 \\ 3 \end{bmatrix},$$

and comparison shows that 5/2 should be the new pivot element, so that x_2 should replace x_3. By the same process as above,

$$\mathbf{M}_2 = \begin{bmatrix} 5/2 & 0 & 0 \\ 1/2 & 1 & 0 \\ 1/2 & 0 & 1 \end{bmatrix}, \quad \mathbf{M}_2^{-1} = \begin{bmatrix} 2/5 & 0 & 0 \\ -1/5 & 1 & 0 \\ -1/5 & 0 & 1 \end{bmatrix},$$

so

$$\mathbf{B}_2^{-1} = \mathbf{M}_2^{-1}\mathbf{M}_1^{-1} = \begin{bmatrix} 2/5 & -3/5 & 0 \\ -1/5 & 4/5 & 0 \\ -1/5 & -1/5 & 1 \end{bmatrix},$$

and

$$\mathbf{h}^{(3)\mathrm{T}} = [-1/5 \quad -1/5 \quad 1] \begin{bmatrix} 1 & 0 \\ 0 & 1 \\ 0 & 0 \end{bmatrix} = [-1/5 \quad -1/5],$$

so that the minimum is reached. The solution is then

$$\mathbf{x}^* = \begin{bmatrix} 2/5 & -3/5 \\ -1/5 & 4/5 \end{bmatrix} \begin{bmatrix} 12 \\ 6 \end{bmatrix} = \begin{bmatrix} 6/5 \\ 12/5 \end{bmatrix}$$

and the solution value is

$$f^* = [-1/5 \quad -1/5]\begin{bmatrix} 12 \\ 6 \end{bmatrix} = -18/5.$$

By tracking back we can show that the solution basic variables are x_2, x_1.

The advantage of this method is that the only storage requirements are for the current $(m+1)\times(m+1)$ inverse matrix $\mathbf{B}_r^{-1}$ and the initial tableau $\mathbf{T}_1$, and that only the required elements are computed.

3.1.6 Degeneracy

A degenerate basic feasible solution is one in which one or more of the basic variables have zero values, so that it is possible to change the basis, without changing the position. There is a famous constructed example, due to Beale (1955), illustrating that under these conditions the simplex method can fail to reach the minimum; you can confirm that this happens for the problem

$$\begin{aligned} &\text{minimise } -3x_1/4 + 150x_2 - x_3/50 + 6x_4 \\ &\text{subject to} \quad x_1/4 - 60x_2 - x_3/25 + 9x_4 \leqslant 0, \quad x_j \geqslant 0, \quad j = 1, 2, 3, 4 \\ &\qquad\qquad\quad x_1/2 - 90x_2 - x_3/50 + 3x_4 \leqslant 0 \\ &\qquad\qquad\qquad\qquad\qquad x_3 \qquad\qquad \leqslant 1 \end{aligned}$$

if the basic variable to drop is always chosen as the first available one in the tableau. A set of six tableaux is produced which then repeat indefinitely, each giving $f = 0$, whereas the true solution is $f = -1/20$.

It is an observed fact that this does not happen in practical applications. However it can be avoided, if an entirely safe procedure is required, by some perturbation procedure. A possible method is outlined in Problem 7.

3.2 Duality

3.2.1 Duality theorem

The symmetric duality theorem of linear programming states that, for the pair of problems

$$\begin{aligned} &\min \mathbf{g}^{\mathrm{T}}\mathbf{x} \text{ subject to } \mathbf{x} \geqslant 0, \mathbf{A}\mathbf{x} \geqslant \mathbf{b} \\ &\max \boldsymbol{\lambda}^{\mathrm{T}}\mathbf{b} \text{ subject to } \boldsymbol{\lambda} \geqslant 0, \boldsymbol{\lambda}^{\mathrm{T}}\mathbf{A} \leqslant \mathbf{g}^{\mathrm{T}} \end{aligned}$$

(1) if either has an unbounded optimum, the other has no feasible solution;
(2) if either has an optimum solution, so has the other, and the optimum values are the same.

The duality theorem of linear programming is a special case of the symmetric duality relation (B) given in Section 1.3, though it requires a separate proof. We dealt there with a pair of problems

Primal $\min K(\mathbf{x}, \boldsymbol{\lambda}) - \boldsymbol{\lambda}^{\mathrm{T}}\nabla_{\lambda}K$ subject to $\mathbf{x} \geq 0$, $\boldsymbol{\lambda} \geq 0$, $\nabla_{\lambda}K \leq 0$

Dual $\max K(\mathbf{x}, \boldsymbol{\lambda}) - \mathbf{x}^{\mathrm{T}}\nabla_{x}K$ subject to $\mathbf{x} \geq 0$, $\boldsymbol{\lambda} \geq 0$, $\nabla_{x}K \geq 0$.

Consider the case of the Lagrangian

$$L(\mathbf{x}, \boldsymbol{\lambda}) = \mathbf{g}^{\mathrm{T}}\mathbf{x} + \boldsymbol{\lambda}^{\mathrm{T}}\mathbf{b} - \boldsymbol{\lambda}^{\mathrm{T}}\mathbf{A}\mathbf{x}. \tag{3.17}$$

Then the primal problem is

$$\min \mathbf{g}^{\mathrm{T}}\mathbf{x} \text{ subject to } \mathbf{x} \geq 0, \mathbf{A}\mathbf{x} \geq \mathbf{b} \tag{3.18}$$

and the dual problem is

$$\max \boldsymbol{\lambda}^{\mathrm{T}}\mathbf{b} \text{ subject to } \boldsymbol{\lambda} \geq 0, \boldsymbol{\lambda}^{\mathrm{T}}\mathbf{A} \leq \mathbf{g}^{\mathrm{T}}. \tag{3.19}$$

The argument given in Section 1.3 shows that

$$\min \mathbf{g}^{\mathrm{T}}\mathbf{x} \geq \max \boldsymbol{\lambda}^{\mathrm{T}}\mathbf{b}.$$

This follows directly from (3.18), (3.19), which show that for any feasible $\boldsymbol{\lambda}$, $\mathbf{x}$,

$$\mathbf{g}^{\mathrm{T}}\mathbf{x} \geq \boldsymbol{\lambda}^{\mathrm{T}}\mathbf{A}\mathbf{x} \geq \boldsymbol{\lambda}^{\mathrm{T}}\mathbf{b} \tag{3.20}$$

and in consequence if either the primal or the dual problem has an unbounded optimum, the other has no feasible solution. This is the first part of the duality theorem. The second part states that if either has a solution, the other has also and the two optimum values are equal. We prove this by a constructive argument similar to that in Section 1.3, that is we construct a solution to the dual for which the value of $\boldsymbol{\lambda}^{\mathrm{T}}\mathbf{b}$ is the same as the minimum f^* of $f = \mathbf{g}^{\mathrm{T}}\mathbf{x}$; this must be the optimum dual solution.

The construction is in terms of the simplex tableau. By the argument used in Section 3.1.5, the final simplex tableau is obtained from the matrix of the initial equations by multiplying by an inverse matrix $(\mathbf{B}^*)^{-1}$. The initial tableau for the primal problem here is

$$\mathbf{T}_1 = \left[\begin{array}{c|c|c|c} \mathbf{A} & -\mathbf{I}_m & 0 & \mathbf{b} \\ \hline -\mathbf{g}^{\mathrm{T}} & 0 & 1 & 0 \end{array}\right] \tag{3.21}$$

and the final one is

$$\mathbf{T}^* = \left[\begin{array}{c|c|c|c} \mathbf{D} & \mathbf{E} & 0 & \mathbf{x}_b^* \\ \hline -\mathbf{d}^{\mathrm{T}} & -\mathbf{e}^{\mathrm{T}} & 1 & f^* \end{array}\right] \quad \text{with } \mathbf{d}, \mathbf{e}, \mathbf{x}_b^* \geqslant 0. \tag{3.22}$$

Then

$$(\mathbf{B}^*)^{-1}\mathbf{T}_1 = \mathbf{T}^* \quad \text{or} \quad \mathbf{T}_1 = \mathbf{B}^*\mathbf{T}^* \tag{3.23}$$

and $\mathbf{B}^*$ is made up of columns of $\mathbf{T}_1$ corresponding to the basic variables and to f. Hence if

$$\mathbf{B}^* = \left[\begin{array}{c|c} \mathbf{R} & 0 \\ \hline -\mathbf{g}_b^{\mathrm{T}} & 1 \end{array}\right], \tag{3.24}$$

where $\mathbf{R}$ is nonsingular, then from (3.21), (3.22), (3.23), (3.24)

$$\left[\begin{array}{c|c} \mathbf{R} & 0 \\ \hline -\mathbf{g}_b^{\mathrm{T}} & 1 \end{array}\right]\left[\begin{array}{c|c|c|c} \mathbf{D} & \mathbf{E} & 0 & \mathbf{x}_b^* \\ \hline -\mathbf{d}^{\mathrm{T}} & -\mathbf{e}^{\mathrm{T}} & 1 & f^* \end{array}\right] = \left[\begin{array}{c|c|c|c} \mathbf{A} & -\mathbf{I} & 0 & \mathbf{b} \\ \hline -\mathbf{g}^{\mathrm{T}} & 0 & 1 & 0 \end{array}\right], \tag{3.25}$$

and so

$$\begin{array}{ll} \text{(i) } \mathbf{A} = \mathbf{RD} & \text{(iv) } -\mathbf{g}^{\mathrm{T}} = -\mathbf{g}_b^{\mathrm{T}}\mathbf{D} - \mathbf{d}^{\mathrm{T}} \\ \text{(ii) } -\mathbf{I} = \mathbf{RE} & \text{(v) } 0 = -\mathbf{g}_b^{\mathrm{T}}\mathbf{E} - \mathbf{e}^{\mathrm{T}} \\ \text{(iii) } \mathbf{b} = \mathbf{R}\mathbf{x}_b^* & \text{(vi) } 0 = -\mathbf{g}_b^{\mathrm{T}}\mathbf{x}_b^* + f^*. \end{array} \tag{3.26}$$

We now show that the vector

$$\boldsymbol{\lambda}^* = \mathbf{e}$$

is a feasible solution to the dual problem and that

$$\boldsymbol{\lambda}^{*\mathrm{T}}\mathbf{b} = f^* = \mathbf{g}^{\mathrm{T}}\mathbf{x}^* \tag{3.27}$$

Proof Using (3.26(v)),

$$\begin{aligned} \mathbf{e}^{\mathrm{T}}\mathbf{A} &= -\mathbf{g}_b^{\mathrm{T}}\mathbf{E}\mathbf{A} \\ &= +\mathbf{g}_b^{\mathrm{T}}\mathbf{R}^{-1}\mathbf{A} \quad &&\text{(from (3.26(ii)))} \\ &= \mathbf{g}_b^{\mathrm{T}}\mathbf{D} &&\text{(from (3.26(i)))} \\ &= \mathbf{g}^{\mathrm{T}} - \mathbf{d}^{\mathrm{T}} &&\text{(from (3.26(iv)))} \\ &\leqslant \mathbf{g}^{\mathrm{T}} &&\text{(since } \mathbf{d}^{\mathrm{T}} \geqslant 0). \end{aligned}$$

Also $\mathbf{e} \geqslant 0$.

Hence $\boldsymbol{\lambda}^* = \mathbf{e}$ is a feasible solution for the dual.

Using (3.26(vi)),

$$
\begin{aligned}
f^* &= \mathbf{g}_b^{\mathrm{T}}\mathbf{x}_b^* \\
&= \mathbf{g}_b^{\mathrm{T}}\mathbf{R}^{-1}\mathbf{b} \quad \text{(from (3.26(iii)))} \\
&= -\mathbf{g}_b^{\mathrm{T}}\mathbf{E}\mathbf{b} \quad \text{(from (3.25(ii)))} \\
&= \mathbf{e}^{\mathrm{T}}\mathbf{b} \quad \text{(from (3.26(v)))} \\
&= \boldsymbol{\lambda}^{*\mathrm{T}}\mathbf{b},
\end{aligned}
$$

as required.

The final simplex tableau therefore contains the solutions $\mathbf{x}^*$, $\boldsymbol{\lambda}^*$ to both the primal and dual problems, a very useful property. Also, from the proof above, we have that

$$\mathbf{d}^{\mathrm{T}} = \mathbf{g}^{\mathrm{T}} - \mathbf{e}^{\mathrm{T}}\mathbf{A} = \mathbf{g}^{\mathrm{T}} - \boldsymbol{\lambda}^{*\mathrm{T}}\mathbf{A}, \tag{3.28}$$

so that the bottom line of the tableau contains also the slack variables for the dual problem.

Example 3.7 Revert again to Problem 3.4.

$$
\begin{aligned}
&\text{(Primal)} \quad \text{minimise } -x_1 - x_2 \quad \text{subject to} \\
&3x_1 + 4x_2 \leqslant 12 \quad x_1, x_2 \geqslant 0 \\
&2x_1 + x_2 \leqslant 6.
\end{aligned}
$$

This in the standard form becomes

$$
\begin{aligned}
&\text{minimise } -x_1 - x_2 \quad \text{subject to} \\
&-3x_1 - 4x_2 \geqslant -12 \quad x_1, x_2 \geqslant 0 \\
&-2x_1 - x_2 \geqslant -6
\end{aligned}
$$

and has as its dual

$$
\begin{aligned}
&\text{maximise } -12\lambda_1 - 6\lambda_2 \text{ subject to} \\
&-3\lambda_1 - 2\lambda_2 \leqslant -1 \\
&-4\lambda_1 - \lambda_2 \leqslant -1.
\end{aligned}
$$

The solution tableau for the primal is, as shown in Table 3.9, as before, so that the primal solution is $x_1^* = 12/5$, $x_2^* = 6/5$, $x_3^* = 0$, $x_4^* = 0$, $f^* = -18/5$. By inspection the dual solution is $\lambda_1^* = 1/5$,

Table 3.9

x_2	0	1	2/5	−3/5	6/5
x_1	1	0	−1/5	4/5	12/5
	0	0	−1/5	−1/5	−18/5

$\lambda_2^* = 1/5$, the dual slack variables are zero, and the dual optimum $-12\lambda_1^* - 6\lambda_2^* = -18/5$.

3.2.2 Lagrange multipliers

The dual variables can be identified with the Lagrange multipliers, and so $\lambda_i^* = 0$ when the i-th inequality is not binding, that is when the slack variable is nonzero, and $\lambda_i^* > 0$ implies that the slack variable is zero. Since the dual of the dual is the primal, the x_j^* are Lagrange multipliers for the dual problem, and hence $x_j^* = 0$ when the j-th dual inequality has a nonzero slack variable, and so on. These results can be combined into the principle of complementary slackness:

$$(\mathbf{g}^{\mathrm{T}} - \boldsymbol{\lambda}^{*\mathrm{T}}\mathbf{A})\mathbf{x}^* = 0, \quad \boldsymbol{\lambda}^{*\mathrm{T}}(\mathbf{A}\mathbf{x}^* - \mathbf{b}) = 0, \tag{3.29}$$

which means that, since every component of each vector is non-negative, one member of each product must be zero. (This also follows from (3.20) and (3.27)).

Also, from (3.18), (3.19) and (3.27), the Lagrangian

$$L(\mathbf{x}, \boldsymbol{\lambda}) = \mathbf{g}^{\mathrm{T}}\mathbf{x} - \boldsymbol{\lambda}^{\mathrm{T}}(\mathbf{A}\mathbf{x} - \mathbf{b}) \tag{3.30}$$

has the saddle point behaviour

$$L(\mathbf{x}^*, \boldsymbol{\lambda}) \leqslant L(\mathbf{x}^*, \boldsymbol{\lambda}^*) \leqslant L(\mathbf{x}, \boldsymbol{\lambda}^*) \tag{3.31}$$

and

$$L(\mathbf{x}^*, \boldsymbol{\lambda}^*) = \mathbf{g}^{\mathrm{T}}\mathbf{x}^* = \boldsymbol{\lambda}^{*\mathrm{T}}\mathbf{b}. \tag{3.32}$$

The dual variable corresponding to an equality constraint is unrestricted in sign, and similarly to any unrestricted primal variable corresponds a dual equality. (These follow from the symmetric duality theorem by converting each equality constraint, and each unrestricted variable, into standard form as in Section 3.1.1.)

Note also that the dual variables represent the sensitivity of the optimum value to the corresponding constraint, that is, as proved generally in Section 1.2.9,

$$\frac{\partial f^*}{\partial b_i} = \boldsymbol{\lambda}_i^*.$$

Here the relation is linear and the connection direct, because as we have shown

$$\begin{aligned} f^* &= g_1 x_1^* + g_2 x_2^* + \cdots + g_n x_n^* \\ &= b_1 \boldsymbol{\lambda}_1^* + b_2 \boldsymbol{\lambda}_2^* + \cdots + b_m \boldsymbol{\lambda}_m^*. \end{aligned} \tag{3.33}$$

3.2.3 Dual simplex method

The normal simplex method generates feasible solutions to the primal, with decreasing values of f, ending with a tableau which contains solutions to both primal and dual and their common optimum value. It is also possible to have a dual simplex method which proceeds through feasible solutions of the dual, increasing the dual objective. A tableau in this will have a bottom row entirely non-positive but a column $\mathbf{x}_b$ containing both positive and negative values. The rule for changing to the next tableau is then

(1) choose the smallest element (a negative element) in $\mathbf{x}_b$, and take this as defining the basic variable x_i which is to leave the basis;
(2) consider *negative* elements in the row $\mathbf{s}_i$, and the j which makes h_j/s_{ij} the smallest numerically;
(3) pivot on s_{ij} as before.

These steps are repeated until all elements in $\mathbf{x}_b$ are non-negative, and this tableau corresponds to the optimum solution to both primal and dual problems.

3.2.4 Sensitivity and duality

In practical situations a very common requirement is for sensitivity analysis, that is to investigate how the optimum solution would change if the original conditions were modified. This can be done using the final simplex tableau, without needing to solve the modified problem from the beginning. The changes may be

(a) in b_i, the constraint limits;
(b) in g_j, the function coefficients;
(c) new variables;
(d) new constraints.

With changes (a) and (b) we need to consider whether the existing basis still provides a feasible solution to both primal and dual problems. Thus for a change of type (a), we require, for the primal (3.21), with basis (3.24)

$$\text{for the new } \mathbf{b}, \mathbf{b}=\mathbf{b}_\nu, \quad \mathbf{R}^{-1}\mathbf{b}_\nu \geqslant 0. \tag{3.34}$$

$$\text{or, from (3.26(ii))}, \quad \mathbf{E}\mathbf{b}_\nu \leqslant 0.$$

If this is true, then the new f^* is found from (3.33); if not, a modified tableau with $-\mathbf{E}\mathbf{b}_\nu$ as its last column and $-\mathbf{g}_b^{\mathrm{T}}\mathbf{E}\mathbf{b}_\nu$ for f can be treated by the dual simplex method to find the new basis and new optimum. Similarly a new $\mathbf{g}$, $\mathbf{g}_\nu$, produces a feasible dual solution if $\mathbf{g}_{b\nu}^{\mathrm{T}}\mathbf{E} \leqslant 0$, $\mathbf{g}_{b\nu}^{\mathrm{T}}\mathbf{D} - \mathbf{g}_\nu^{\mathrm{T}} \leqslant 0$ from (3.26(iv), (v)). If this is so, the new f^* is found from (3.33), otherwise the tableau is used with the normal simplex method starting with the values found in the f row and a value $-\mathbf{g}_{b\nu}^{\mathrm{T}}\mathbf{E}\mathbf{b}$ for f.

Changes (c), (d) can be treated similarly. The introduction of a new variable, x_ν, means a new dual constraint

$$a_{1\nu}\lambda_1 + a_{2\nu}\lambda_2 + \cdots + a_{m\nu}\lambda_m \leqslant g_\nu. \tag{3.35}$$

If this is satisfied by $\boldsymbol{\lambda}^*$, then the existing solution is feasible and so optimal. If not, there is a positive h_ν,

$$h_\nu = a_{1\nu}\lambda_1^* + a_{2\nu}\lambda_2^* + \cdots + a_{m\nu}\lambda_m^* - g_\nu. \tag{3.36}$$

The tableau is completed by finding $s_{i\nu}$. Writing $\mathbf{s}^\nu$ for the column vector $\{s_{i\nu}\}$, we have

$$\mathbf{R}\mathbf{s}^\nu = \mathbf{a}^\nu$$

and so

$$\mathbf{s}^\nu = -\mathbf{E}\mathbf{a}^\nu. \tag{3.37}$$

Similarly a new constraint

$$a_{\nu 1}x_1 + a_{\nu 2}x_2 + \cdots + a_{\nu n}x_n \geqslant b_\nu, \tag{3.38}$$

if not satisfied by the existing solution, produces a negative element in the final column:

$$x_\nu = a_{\nu 1}x_1^* + a_{\nu 2}x_2^* + \cdots + a_{\nu n}x_n^* - b_\nu,$$

and elements in the row $\mathbf{s}_\nu$ calculated as for (b) above.

Example 3.8 Given the original problem,

minimise $f = -3x_1 - x_2 + 2x_3$ subject to

$$2x_1 + 3x_2 - x_3 \leqslant 24$$
$$-2x_1 + 2x_2 + 5x_3 \leqslant 20$$
$$x_1 + x_2 - 2x_3 \leqslant 40 \quad x_j \geqslant 0, \quad j = 1, 2, \ldots, 6,$$

the usual simplex method produces a final tableau, including slack variables, as shown in Table 3.10, so that $x_1^* = 35/2$, $x_2^* = 0$, $x_3^* = 11$, $f^* = -149/2$.

Table 3.10

x_1	1	17/8	0	5/8	1/8	0	35/2
x_3	0	5/4	1	1/4	1/4	0	11
x_6	0	11/8	0	−1/8	3/8	1	89/2
	0	−63/8	0	−19/8	−7/8	0	−149/2

The dual problem is

$$\begin{aligned}\text{maximise}\quad & -24\lambda_1-20\lambda_2-40\lambda_3\\ \text{subject to}\quad & -2\lambda_1+2\lambda_2-\lambda_3\leqslant -3\\ & -3\lambda_1-2\lambda_2-\lambda_3\leqslant -1\\ & \lambda_1-5\lambda_2+2\lambda_3\leqslant -2,\quad \lambda_i\geqslant 0\end{aligned}$$

and the solution is $\lambda_1^*=19/8$, $\lambda_2^*=7/8$, $\lambda_3^*=0$. A new constraint

$$-2x_1-3x_2+4x_3\leqslant 6$$

is now introduced. This is not satisfied by $\mathbf{x}^*$, and the slack variable, x_7, is -3. The amended tableau acquires an additional row corresponding to x_7, and the j-th row element is $-\mathbf{g}_b^{\mathrm{T}}\mathbf{s}^j+g_j$. Thus

$$\begin{aligned} s_{42}&=-(-2)17/8-(4)5/4+(-3)=-15/4\\ s_{44}&=-(-2)5/8-(4)1/4 \qquad\qquad = \quad 1/4\\ s_{45}&=-(-2)1/8-(4)1/4 \qquad\qquad = -3/4\end{aligned}$$

and the new tableau is Table 3.11.

Table 3.11

x_1	1	17/8	0	5/8	1/8	0	0	35/2
x_3	0	5/4	1	1/4	1/4	0	0	11
x_6	0	11/8	0	−1/8	3/8	1	0	89/2
x_7	0	−15/4	0	1/4	(−3/4)	0	1	−3
	0	−63/8	0	−19/8	−7/8	0	0	−149/2

Operating the dual simplex procedure, x_7 must leave the basis, a negative s must be chosen as the pivot, and since $(63/8)/(15/4)>(7/8)/(3/4)$ the pivot has to be $-3/4$. Pivoting as usual produces Table 3.12, giving an amended solution $f=-71$, $x_1=17$, $x_3=10$, which is optimal.

Table 3.12

x_1								17
x_3								10
x_6								43
x_5								4
	0	−7/2	0	−8/3	0	0	−7/6	−71

Problems

(1) Minimise $f=-2x_1-x_2$ subject to

$$\begin{aligned} x_1+4x_2&\leqslant 24 \\ x_1+2x_2&\leqslant 14 \quad x_1, x_2\geqslant 0 \\ 2x_1-\ x_2&\leqslant 8 \\ x_1-\ x_2&\leqslant 3. \end{aligned}$$

(a) Solve graphically, (b) solve by simplex method, (c) list all basic feasible solutions.

(2) Minimise $f=-2x_1-3x_2-5x_3$ subject to

$$\begin{aligned} 3x_1+10x_2+5x_3&\leqslant 15 \\ 33x_1-10x_2+9x_3&\leqslant 33 \quad x_1, x_2, x_3\geqslant 0 \\ x_1+\ 2x_2+\ x_3&\geqslant 4 \end{aligned}$$

(3) Minimise $f=-4x_1-x_2-3x_3-5x_4$ subject to

$$\begin{aligned} -4x_1+6x_2+5x_3-4x_4&\leqslant 20 \\ 3x_1-2x_2+4x_3+\ x_4&\leqslant 10 \quad x_1, x_2, x_3, x_4\geqslant 0 \\ 8x_1-3x_2+3x_3+2x_4&\leqslant 20. \end{aligned}$$

(4) For a certain LP problem, the objective function has a finite minimum f_0. The k-th stage of the simplex method gives a degenerate basic feasible solution with exactly one basic variable taking the value zero, and the objective function value $f_k>f_0$. Prove that this k-th basis cannot recur in subsequent stages.

(5) Verify that a degenerate basic feasible solution of the primal providing a minimum f^* for the primal objective can imply multiple solutions for the dual problem. Illustrate by considering the primal problem

$$\begin{aligned} \text{minimise} \quad & f=-5x_1+x_2 \\ \text{subject to} \quad & -\ x_1+x_2\geqslant 0 \\ & -2x_1 \qquad \geqslant -1 \quad x_1, x_2\geqslant 0 \\ & -3x_1-x_2\geqslant -2. \end{aligned}$$

List the multiple solutions to the dual.

(6) Write out the dual of the problem

maximise $f = x_1 + x_2 + x_3$

subject to $2x_1 + x_2 + 2x_3 \leqslant 2 \quad x_1, x_2, x_3 \geqslant 0$

$4x_1 + 2x_2 + x_3 \leqslant 2.$

Solve either problem and, from the solution, pick out the solution of the other.

(7) Problems of cycling due to degeneracy can be avoided by perturbing the original specification. Illustrate this by considering the problem derived from Beale's problem by amending the values b_i by multiples of the columns:

minimise $f = -\frac{3}{4}x_1 + 150x_2 - \frac{1}{50}x_3 + 6x_4$ subject to

$$\tfrac{1}{4}x_1 - 60x_2 - \tfrac{1}{25}x_3 + 9x_4 + x_5 = 0 + \tfrac{1}{4}\varepsilon - 60\varepsilon^2 - \tfrac{1}{25}\varepsilon^3 + 9\varepsilon^4 + \varepsilon^5$$

$$\tfrac{1}{2}x_1 - 90x_2 - \tfrac{1}{50}x_3 + 3x_4 + x_6 = 0 + \tfrac{1}{2}\varepsilon - 90\varepsilon^2 - \tfrac{1}{50}\varepsilon^3 + 3\varepsilon^4 + \varepsilon^6$$

$$x_3 + x_7 = 1 + \varepsilon^3 + \varepsilon^7$$

Assume that ε is a small positive quantity, so that any term in $\varepsilon^r <$ any term in ε^{r-1}, and show that the ambiguity in changing bases is then removed and that the minimum is attained.

(8) Show that direct application of the Kuhn–Tucker conditions to the problem

minimise $f = \mathbf{g}^{\mathrm{T}}\mathbf{x}$ subject to $\mathbf{Ax} - \mathbf{b} = 0$, $\mathbf{x} \geqslant 0$,

gives optimality when there exist $\boldsymbol{\lambda}^*$ and $\boldsymbol{\mu}^* \geqslant 0$ such that

$$\mathbf{g}^{\mathrm{T}} - \boldsymbol{\lambda}^{*\mathrm{T}}\mathbf{A} - \boldsymbol{\mu}^{*\mathrm{T}} = 0$$

$$\boldsymbol{\mu}^* \geqslant 0, \quad \boldsymbol{\mu}^{*\mathrm{T}}\mathbf{x}^* = 0$$

Verify that this means

$$\boldsymbol{\lambda}^{*\mathrm{T}}\mathbf{A}^j \leqslant \mathbf{g}_j \quad \text{when} \quad \mathbf{x}_j^* = 0.$$

$$\boldsymbol{\lambda}^{*\mathrm{T}}\mathbf{A}^j = \mathbf{g}_j \quad \text{when} \quad \mathbf{x}_j^* > 0.$$

(9) Solve graphically to show that the following problem has an unbounded solution. Write down the dual problem and by introducing surplus variables and using the simplex method, verify that it has no feasible solution.

$$\begin{aligned} \text{Maximise} \quad & 3x_1 + 4x_2 \\ \text{subject to} \quad & -x_1 + x_2 \leqslant 1 \\ & x_1 + x_2 \geqslant 4 \quad x_1, x_2 \geqslant 0. \\ & x_1 - 3x_2 \leqslant 3 \end{aligned}$$

(10) Show that if for an optimal solution of an LP problem, some coefficient h_j of a nonbasic variable is zero, then there exists an alternative optimal solution with x_j in the basis.

(11) In seeking to minimise $f = x_1 - 3x_2 + 2x_3$ subject to

$$\begin{aligned} 3x_1 - x_2 + 2x_3 &\leqslant 7 \\ -2x_1 + 4x_2 \qquad &\leqslant 12 \quad x_1, x_2, x_3 \geqslant 0 \\ -4x_1 + 3x_2 + 8x_3 &\leqslant 10, \end{aligned}$$

slack variables are introduced and the final tableau is Table 3.13. Starting from this final tableau, answer the following queries:

(a) If the objective function is changed to $x_1 - 3x_2 - x_3$ is the solution still optimal? If not, find the new optimal solution.
(b) A new nonnegative variable x_7, is introduced into the first two inequalities, and into f; now $f = x_1 - 3x_2 + 2x_3 - 3x_7$, and $3x_1 - x_2 + 2x_3 + 3x_7 \leqslant 7$, $-2x_1 + 4x_2 + 2x_7 \leqslant 12$.

Does this change the solution? If so, find the new one.

Table 3.13

x_1	1	0	4/5	2/5	1/10	0	4
x_2	0	1	2/5	1/5	3/10	0	5
x_6	0	0	10	1	−1/2	1	11
	0	0	−12/5	−1/5	−4/5	0	−11

(12) The optimal solution to

$$\mathbf{Ax} = \mathbf{b}, \quad \mathbf{x} \geqslant 0, \quad f = \mathbf{g}^T\mathbf{x} \text{ to minimise}$$

is $\mathbf{x}^*$, and $\boldsymbol{\lambda}^*$ is the optimal solution to the corresponding dual problem. Prove that, if now $\mathbf{b}$ is replaced by $\mathbf{b}'$, and the new optimal solution to the primal is $\mathbf{x}'^*$, then

$$\mathbf{g}^T(\mathbf{x}^* - \mathbf{x}'^*) \leqslant \boldsymbol{\lambda}^{*T}(\mathbf{b} - \mathbf{b}')$$

(13) The solution of an LP minimising problem $\mathbf{Ax} \geqslant \mathbf{b}$, $\mathbf{x} \geqslant 0$, $\min \mathbf{g}^T\mathbf{x}$, $\mathbf{x}$ an n-vector, has been found by the simplex method. It

is now decided that one of the variables, x_s, can be allowed to become negative. What conditions must hold in the final simplex tableau for this change to improve the solution? Distinguish between the cases where x_s is a "real" variable, i.e. $s \leqslant n$, and where x_s is a slack variable.

(14) Take the following LP problem:

minimise $-x_1+x_2$

subject to $x_1+x_2 \geqslant p \quad x_1, x_2 \geqslant 0$

$qx_1+x_2 \leqslant 10$

as the primal problem. Then

(a) write down the dual problem;
(b) state the values of p, q for which a unique solution exists to the primal;
(c) state the values of p, q for which the solution space is (i) empty (ii) unbounded. What happens to the dual problem in each of these cases?

(15) $\mathbf{A}$ is a given $m \times n$ matrix, $\mathbf{b}$ is a given m-vector. Prove that

(a) if no $\mathbf{x} \geqslant 0$ exists such that $\mathbf{Ax}=\mathbf{b}$, then a $\mathbf{y}$ exists such that $\mathbf{A}^{\mathrm{T}}\mathbf{y} \geqslant 0$ and $\mathbf{b}^{\mathrm{T}}\mathbf{y}<0$;
(b) the systems $\mathbf{Ax}=0$, $\mathbf{x} \geqslant 0$ and $\mathbf{A}^{\mathrm{T}}\mathbf{y} \geqslant 0$ have a solution such that $\mathbf{A}^{\mathrm{T}}\mathbf{y}+\mathbf{x}>0$;
(c) if $\mathbf{A}$ is skew symmetric, then the system

$$\mathbf{Ay} \geqslant 0, \quad \mathbf{y} \geqslant 0, \quad \mathbf{Ax} \geqslant 0, \quad \mathbf{x} \geqslant 0$$

has a solution such that

$$\mathbf{Ay}+\mathbf{x}>0 \quad \text{and} \quad \mathbf{Ax}+\mathbf{y}>0;$$

(d) under the same conditions as in (c),

$$\mathbf{Aw} \geqslant 0, \quad \mathbf{w} \geqslant 0$$

has a solution such that $\mathbf{Aw}+\mathbf{w}>0$.
(*Hint:* These are all consequences of the duality theorem.)

4

Applications of linear programming

Three common applications of LP will be dealt with in this chapter; others are mentioned in the problems. The first is the problem of allocation of resources when the constraints and objective are linear – the dual variables then have an important interpretation as shadow prices. The second is the special class of problem known as transportation, transshipment and assignment, for which simplified algorithms are available. The third application is to simple two-person zero-sum game theory where the primal and dual problems correspond to the separate viewpoints of the two players.

In the final section a brief account is given of separable programming, that is the application of LP methods to problems where the objective and constraints are nonlinear but of a form which can be approximated by piecewise linear functions. An approximate solution only is obtained by this method – accurate methods for minimising under constraints are discussed in the next chapter. Separable programming is of use in cases where the advantages of being able to use LP methods outweigh the disadvantages of approximation.

4.1 Allocation of resources

We state first a typical problem of this type.

Example 4.1 A manufacturer uses materials A, B and C to make products P_1, P_2, P_3, P_4. Table 4.1 gives the amounts of A,

Table 4.1

Product	P_1	P_2	P_3	P_4
Amount of A/unit	5	1	9	12
Amount of B/unit	4	3	4	1
Amount of C/unit	3	2	5	10
Profit/unit	12	5	15	10

B and C required to make one unit of each product, and the profit (£) per unit. He has available 1500 units of A, 1000 units of B, 800 units of C. What quantities of each product should be made so as to maximise his profit?

Writing x_1, x_2, x_3, x_4 as the quantities, the conditions lead directly to the LP problem

$$\mathbf{Ax} \leqslant \mathbf{b}, \mathbf{x} \geqslant 0, \text{maximise } f = \mathbf{c}^{\mathrm{T}}\mathbf{x}, \text{ where}$$

$$\mathbf{A} = \begin{bmatrix} 5 & 1 & 9 & 12 \\ 4 & 3 & 4 & 1 \\ 3 & 2 & 5 & 10 \end{bmatrix}, \quad \mathbf{b} = \begin{bmatrix} 1500 \\ 1000 \\ 800 \end{bmatrix}, \quad \mathbf{c}^{\mathrm{T}} = (12, 5, 15, 10) \quad (4.1)$$

The final tableau proves to be that shown in Table 4.2, (x_5 being the slack variable for the first inequality, corresponding to material A) and so the solution is that the manufacturer should make 225 units of P_1, 25 of P_3, use up all of B and C, and thereby make a profit of £3075. (Note that this is a maximising problem, and so the solution is obtained when all elements in the bottom line are positive; these are the solutions of the dual problem.)

Table 4.2

x_5	0	−9/4	0	−31/4	1	1/4	−2	150
x_1	1	7/8	0	−35/8	0	5/8	−1/2	225
x_3	0	−1/8	1	37/8	0	−3/8	1/2	25
	0	+29/8	0	+55/8	0	+15/8	+3/2	+3075

Additional information comes directly from the dual solution, $\lambda_1^* = 0$, $\lambda_2^* = 15/8$, $\lambda_3^* = 3/2$. Because of the sensitivity relations discussed in Section 3.2, the manufacturer's profit f would increase by λ_2^* per unit increase in the limit of B (i.e. if the 1000 becomes 1001), so that λ_2^* has the dimensions of £/unit of B, and can be termed a "price" for B. Similarly λ_3^* is a price for C, and the price for A is zero. These prices represent the level at which it would pay the manufacturer to buy in extra quantities of materials B and C. Thus if he can buy B at a price p_2 less than £1.875 per unit, or C at p_3 less than £1.5 per unit, then his profit will increase; it would not of course pay him to buy any extra A since he is not using up all his allocation of A anyway.

If the limits on quantities of B and C are increased, by amounts b and c, then, again as in Section 3.2, the new solution is, with the same basis,

$$x_1^* = (5/8)(1000+b) - (1/2)(800+c) = 225 + 5b/8 - c/2$$
$$x_3^* = (-3/8)(1000+b) + (1/2)(800+c) = 25 - 3b/8 + c/2$$
$$f = 3075 + b(15/8 - p_2) + c(3/2 - p_3).$$

This solution holds as long as $x_1^*, x_3^* \geqslant 0$; that is, for

$$\begin{aligned} -5b/8 + c/2 &\leqslant 225 \\ 3b/8 - c/2 &\leqslant 25. \end{aligned} \tag{4.2}$$

If conditions (4.2) are violated, then the basis must be changed and a new solution found.

4.2 Transportation problems

4.2.1 Definitions, general properties

The name "transportation problem" is given to a common class of problem which can arise in the study of optimising transport schedules but occurs in other contexts also. We define a typical problem.

Example 4.2 Material is stocked in given amounts at different sites, with the amount a_i at the i-th site, $i = 1, \ldots, m$. It is to be transported to different destinations, the amount b_j being required at the j-th destination, $j = 1, \ldots, n$, and $\sum_{i=1}^m a_i = \sum_{j=1}^n b_j$. The cost of transporting an amount x_{ij} of material from site i to destination j is $c_{ij}x_{ij}$, where $\{c_{ij}\}$ is a given matrix. What is the cheapest way of transporting the material?

The mathematical formulation is to find x_{ij} such that

$$\left.\begin{aligned} \sum_{j=1}^{n} x_{ij} &= a_i, \quad i = 1, \ldots, m \\ \sum_{i=1}^{m} x_{ij} &= b_j, \quad j = 1, \ldots, n, \end{aligned}\right. \qquad x_{ij} \geqslant 0 \tag{4.3}$$

where

$$\sum_{i=1}^{m} a_i = \sum_{j=1}^{n} b_j \tag{4.4}$$

and to minimise

$$f = \sum_{i=1}^{m} \sum_{j=1}^{n} c_{ij} x_{ij}. \tag{4.5}$$

The relation (4.4) balances output and input and means that the whole quantity is transferred. It can always be satisfied, even when this is not actually the case, by introducing a dummy source or destination, with costs zero, so we take it as part of the typical transportation problem.

The problem defined by (4.3), (4.4), (4.5) is clearly a LP problem and can be solved by the simplex method. However, because of its structure a simpler special algorithm is available, which we can now derive. We develop first certain properties.

(1) The $(m+n)$ equality constraints on x_{ij},

$$\sum_{j=1}^{n} x_{ij} = a_i, \quad \sum_{i=1}^{m} x_{ij} = b_j, \tag{4.6}$$

given relation (4.4) are clearly dependent (by summing them). However, if any one of the a_i or b_j constraints is omitted, the resulting set become independent, since then some x_{ij} occur in one equation only. Hence the rank is $m+n-1$ and this is the number of basic variables.

(2) The dual problem is to find u_i, v_j satisfying

$$u_i + v_j \leqslant c_{ij} \tag{4.7}$$

and maximising

$$g = \sum_{i=1}^{m} a_i u_i + \sum_{j=1}^{n} b_j v_j$$

with u_i, v_j unrestricted in sign. Here we have used u_i for the dual variable corresponding to the a_i equality, and v_j for the b_j equality, as a convenient notation.

(3) Because of complementary slackness (3.29) we know also that

$$u_i + v_j = c_{ij} \quad \text{when } x_{ij} \text{ is basic.} \tag{4.8}$$

Hence the $m+n$ dual variables are related by $m+n-1$ relations, and one dual variable is arbitrary.

(4) Suppose a feasible primal solution has been produced, that is we have $x_{ij}^{(r)}$ satisfying (4.3)

$$\sum_{j=1}^{n} x_{ij}^{(r)} = a_i, \quad \sum_{i=1}^{m} x_{ij}^{(r)} = b_j, \quad x_{ij}^{(r)} \geqslant 0.$$

Further, suppose that $u_i^{(r)}$, $v_j^{(r)}$ have been found satisfying (4.8). If these also satisfy (4.7), then we have found the optimal solution (since feasibility of both primal and dual means optimality). If not, then there exist cells *ij* where (4.7) is violated. We now produce a form equivalent to the simplex method which specifies how the primal solution should be modified. Multiplying each equation in (4.3) by its dual variable and adding gives

$$\sum_{i,j\in K} (u_i^{(r)}x_{ij}^{(r)} + v_j^{(r)}x_{ij}^{(r)}) = \sum_{i,j} (a_i u_i^{(r)} + b_j v_j^{(r)}), \tag{4.9}$$

where the set K is the set of indices such that $x_{ij}^{(r)}$ is a basic variable; and since (4.8) holds for $i, j \in K$, then

$$\sum_{i,j\in K} (u_i^{(r)}x_{ij}^{(r)} + v_j^{(r)}x_{ij}^{(r)}) = f^{(r)}, \tag{4.10}$$

the cost of the r-th solution. Now suppose an additional route pq is considered so that x_{ij} is basic in the set $K_+ = K + (pq)$ and is subject still to conditions (4.6). The new cost is

$$\begin{aligned} f &= \sum_{i,j\in K} c_{ij}x_{ij} + c_{pq}x_{pq} \\ &= \sum_{i,j\in K} (u_i^{(r)} + v_j^{(r)})x_{ij} + c_{pq}x_{pq} \\ &= \sum_{i,j\in K_+} (u_i^{(r)} + v_j^{(r)})x_{ij} + (c_{pq} - u_p^{(r)} - v_q^{(r)})x_{pq}. \end{aligned}$$

Since the primal is still feasible, the first sum is

$$\sum_{i,j} u_i^{(r)}a_i + v_j^{(r)}b_j = f^{(r)} \quad \text{(from (4.10))}.$$

Hence

$$f = f^{(r)} + (c_{pq} - u_p^{(r)} - v_q^{(r)})x_{pq}. \tag{4.11}$$

Thus if the new cell pq is chosen from cells where the dual solution is infeasible, then making x basic there will decrease f. It is conventional, as in the simplex method, to choose for pq the cell where the infeasibility $(u_p + v_q - c_{pq})$ is greatest.

4.2.2 Transportation algorithm

The method of solution then contains the following steps:
Step 1 Find a basic feasible solution to the primal, containing $m+n-1$ non-negative x_{ij}.

Step 2 Find u, v, satisfying (4.8) in the basic cells.
Step 3 Check whether (4.7) is satisfied in all cells. If it is, the solution is reached; if not, find where the new basic variable should be.
Step 4 Obtain a new basic feasible solution to the primal including this and dropping one of the existing basic variables. Repeat from Step 2.

We now specify how each of these steps is carried out. We prove first a convenient property of eqs. (4.6).

The equations are triangular in any set of feasible basic variables, that is at least one equation contains only one basic variable.

Proof Suppose each equation contains two or more basic variables. Since these are in equations of form $\sum_j x_{ij} = a_i$, and the x_{ij} occurring in each equality are distinct, the number K of basic variables must satisfy

$$K \geqslant 2m.$$

Similarly

$$K \geqslant 2n,$$

and so

$$K \geqslant m + n.$$

But this is a contradiction since we have already shown that $K = m + n - 1$. Hence the result follows.

This means, once we have decided the set of $(m+n-1)$ x_{ij} to be basic, the equations we have to solve to find their values contain at least one of the form

$$x_{ij} = a_i \quad (\text{or } b_j).$$

Dropping this equation from the set, and substituting the value a_i into the only other equation involving x_{ij} produces a transportation problem with effectively one less source, and so this is also triangular in the remaining basic variables. Continuing this process means that the basic variables can be evaluated one by one, by processes of addition and subtraction only. It is the substitution of these simpler processes for the matrix inversion implicit in the simplex method which makes the transportation problem so much easier to solve. A further consequence is that if the totals a_i, b_j are integers, then the solution x_{ij} is also integral.

Solution of example The application of these results can now be illustrated on a particular example.

Example 4.3 Solve the transportation problem with

$$m = 3, \quad n = 4, \quad \mathbf{a}^T = (4, 3, 5), \quad \mathbf{b}^T = (2, 1, 5, 4),$$

and cost matrix

$$\mathbf{C} = \begin{bmatrix} 5 & 4 & 9 & 2 \\ 1 & 6 & 3 & 7 \\ 4 & 9 & 2 & 8 \end{bmatrix}.$$

Step 1 Find a basic feasible solution for the primal – that is, a solution with 6 (=4+3−1) basic non-negative variables. We know as above that these can be determined one by one. So choose x_{11} as a basic variable (or any other – this is the simplest choice and corresponds to what is commonly called the North-West corner rule, for obvious reasons!).

The two equations involving x_{11} here are

$$x_{11} + x_{12} + x_{13} + x_{14} = 4 \tag{4.12}$$

and

$$x_{11} + x_{21} + x_{31} = 2. \tag{4.13}$$

Clearly we can satisfy these feasibly by taking $x_{11} = 2$, $x_{21} = 0$, $x_{31} = 0$, thus removing eq. (4.13), and replacing (4.12) by

$$x_{12} + x_{13} + x_{14} = 2. \tag{4.14}$$

The new set of equations contains 9 variables and represents a new transportation problem with $\mathbf{a}^T = (2, 3, 5)$ and $\mathbf{b}^T = (1, 5, 4)$; there will be five basic variables in this solution. The process can now be repeated, and, since one equation is removed completely each time, a solution will be reached in six steps, the final equation being automatically satisfied. The solution is

$$\{x_{ij}\}^{(1)} = \begin{bmatrix} 2 & 1 & 1 & 0 \\ 0 & 0 & 3 & 0 \\ 0 & 0 & 1 & 4 \end{bmatrix}$$

with a cost $f^{(1)} = 10 + 4 + 9 + 9 + 2 + 32 = 66$.

Step 2 We then solve (4.8) for u_i, v_j; that is, we define

$$\left.\begin{array}{llllll} u_1 & +v_1 & & & & =5 \\ u_1 & & +v_2 & & & =4 \\ u_1 & & & +v_3 & & =9 \\ u_2 & & & +v_3 & & =3 \\ u_3 & & & +v_3 & & =2 \\ u_3 & & & & +v_4 & =8 \end{array}\right\} \quad (4.15)$$

and taking $u_1^{(1)}=1$ arbitrarily, these give successively

$$u_2^{(1)}=-5, \quad u_3^{(1)}=-6, \quad v_1^{(1)}=4, \quad v_2^{(1)}=3, \quad v_3^{(1)}=8, \quad v_4^{(1)}=14.$$

Step 3 The inequality (4.7) is violated in two cells

$$u_1^{(1)}+v_4^{(1)}=15>2$$

and

$$u_2^{(1)}+v_4^{(1)}=\ 9>7.$$

We choose to include x_{14} as a new basic variable.
Step 4 Consider the equations connecting x_{14} with the existing basic variables, keeping all other nonbasic variables zero. Writing these out in full,

$$\left.\begin{array}{l} x_{11}+x_{12}+x_{13}+x_{14} = 4 \\ x_{23} = 3 \\ x_{33}+x_{34} = 5 \\ x_{11} = 2 \\ x_{12} = 1 \\ x_{13}+x_{23}+x_{33} = 5 \\ x_{14}+x_{34} = 4 \end{array}\right\} \quad (4.16)$$

Reducing these by eliminating equations which define variables, e.g. $x_{23}=3$, produces a set

$$\begin{array}{l} x_{13}+x_{14} = 1 \\ x_{33}+x_{34}=5 \\ x_{13}+x_{33} = 2 \\ x_{14}+x_{34}=4, \end{array} \quad (4.17)$$

for which the solution is

$$x_{13}=1-x_{14}, \quad x_{33}=1+x_{14}, \quad x_{34}=4-x_{14}.$$

Hence the new basic feasible solution is found by making x_{13} nonbasic and is

$$x_{14}^{(2)}=1, \quad x_{33}^{(2)}=2, \quad x_{34}^{(2)}=3.$$

This is the process corresponding to the pivot step in the simplex method. It can be visualised very simply in two dimensions; the insertion of a positive term, θ, into the cell 14 while keeping row and column totals the same involves a sequence of $\pm\theta$ operations on a chain of basic variables, $x_{ij}^{(1)}$, as shown.

$$\begin{bmatrix} 2 & 1 & 1-\theta & 0+\theta \\ 0 & 0 & 3 & 0 \\ 0 & 0 & 1+\theta & 4-\theta \end{bmatrix}.$$

The new solution is then

$$\{x_{ij}^{(2)}\}=\begin{bmatrix} 2 & 1 & 0 & 1 \\ 0 & 0 & 3 & 0 \\ 0 & 0 & 2 & 3 \end{bmatrix},$$

for which $f^{(2)}=10+4+2+9+4+24=53$, a change of $13=(15-2)\times 1$.

The process is repeated from step 2, and a final tableau is

$$\{x_{ij}^{*}\}=\begin{bmatrix} 0 & 0 & 0 & 4 \\ 2 & 1 & 0 & 0^* \\ 0 & 0 & 5 & 0^* \end{bmatrix},$$

with a cost $f^*=26$. There are two basic zeros, shown starred, so this is a degenerate solution. The corresponding dual solution is

$$u_1^*=1, \quad u_2^*=6, \quad u_3^*=7, \quad v_1^*=-5, \quad v_2^*=0, \quad v_3^*=-5, \quad v_4^*=1$$

and you can verify that this is feasible.

If the cost is decreased at each stage, the process must terminate. Degeneracy means a change of basis without change of cost, but in general gives no difficulty.

4.2.3 Transshipment problems

A slightly more general form of the transportation problem arises when there are intermediate depots which receive and despatch material. The generalised form of Example 4.3 is given.

Example 4.4 Material is stocked in given amounts a_i at sites P_i, $i \in I_1$; it is required in amounts b_i at sites P_i, $i \in I_2$; it can pass through sites P_i, $i \in I_3$

$$\sum_{i \in I_1} a_i = \sum_{i \in I_2} b_i.$$

Communication is possible between any pair of sites with cost $c_{ij}x_{ij}$ for transporting quantity x_{ij} from site i to site j. What is the cheapest way of transporting the material?

Relations (4.3) become in this case

$$\left.\begin{aligned} \sum_{j \neq i} x_{ij} - \sum_{j \neq i} x_{ji} &= a_i \qquad i \in I_1 \\ \sum_{j \neq i} x_{ij} - \sum_{j \neq i} x_{ji} &= -b_i \quad\; i \in I_2 \\ \sum_{j \neq i} x_{ij} - \sum_{j \neq i} x_{ji} &= 0 \qquad\; i \in I_3 \\ x_{ij} &\geqslant 0, \end{aligned}\right\} \tag{4.18}$$

and we seek to minimise $f = \sum_{j \neq i} c_{ij}x_{ij} + c_{ji}x_{ji}$. (4.19)

Note that we can have $c_{ij} \neq c_{ji}$, but that actual motion will only be one way, that is $x_{ij} > 0$ means $x_{ji} = 0$.

The relations (4.18) can be put into the form (4.3) by introducing $c_{ii} = 0$, and a flow x_{ii}, together with a buffer stock B, a large number; then

$$\begin{aligned} x_{ii} + \sum_{j \neq i} x_{ij} &= B + a_i, & x_{ii} + \sum_{j \neq i} x_{ji} &= B & i \in I_1 \\ x_{ii} + \sum_{j \neq i} x_{ij} &= B, & x_{ii} + \sum_{j \neq i} x_{ji} &= B + b_i & i \in I_2 \\ x_{ii} + \sum_{j \neq i} x_{ij} &= B, & x_{ii} + \sum_{j \neq i} x_{ji} &= B & i \in I_3. \end{aligned} \tag{4.20}$$

You can verify that this formulation enables the transportation algorithm to be used, and that no B occurs in the final cost – it is analogous to the M method for Phase 1 of the simplex method (3.1.4).

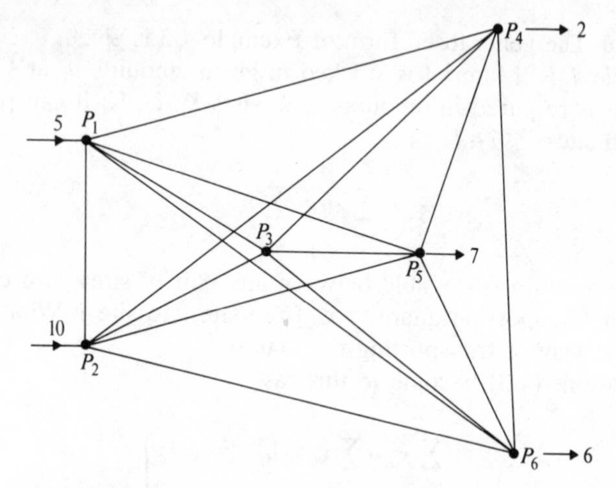

Fig. 4.1. Network for Example 4.5.

We illustrate with a simple example:

Example 4.5 Solve the transshipment problem of Example 4.4 with the network illustrated in Fig. 4.1 and costs as shown in Table 4.3.

Table 4.3

Cost c_{ij} From \ To	P_1	P_2	P_3	P_4	P_5	P_6
P_1	0	1	2	5	7	10
	1	0	6	5	9	5
P_3	2	6	0	2	1	3
P_4	5	5	2	0	1	3
P_5	7	9	1	1	0	1
P_6	10	5	3	3	1	0

$$a_1 = 5, \quad a_2 = 10, \quad b_4 = 2, \quad b_5 = 7, \quad b_6 = 6.$$

You can confirm that an optimal solution is as illustrated in Fig. 4.2. It is

$$x_{13}^* = 7, \quad x_{21}^* = 2, \quad x_{24}^* = 2, \quad x_{26}^* = 6, \quad x_{35}^* = 7$$

(the "real" flows) together with

$$x_{11}^* = B - 2, \quad x_{22}^* = B, \quad x_{33}^* = B - 7, \quad x_{44}^* = x_{55}^* = x_{66}^* = B.$$

The dual solution is

$$\mathbf{u}^{*\mathrm{T}} = (1, 2, -1, -3, -2, -3), \quad \mathbf{v}^{*\mathrm{T}} = (-1, -2, 1, 3, 2, 3)$$

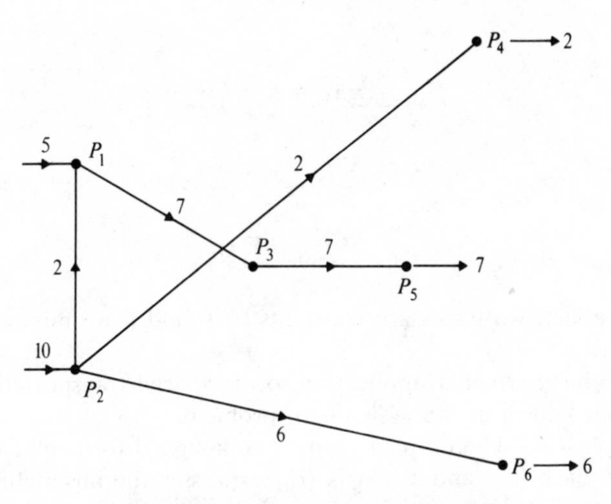

Fig. 4.2. Solution of Example 4.5.

and since $u_i + v_j = c_{ij}$ in the nonbasic cells 34 and 54, there are other solutions with the same value, $f^* = 63$.

This form of problem also occurs in contexts other than transport: see Problem 11.

4.2.4 Assignment problems

Before discussing these we give first a useful property of all transportation-type problems.

Under conditions

$$\sum_{j=1}^{n} x_{ij} = a_i, \quad \sum_{i=1}^{m} x_{ij} = b_j,$$

the two problems

(1) minimise $f = \sum\sum c_{ij} x_{ij}$ (4.21)

and

(2) minimise $f' = \sum\sum c'_{ij} x_{ij}$,

where $c'_{ij} = c_{ij} - K_i - l_j$, and K_i, l_j are constants, have the same solution x^*_{ij}.

Proof

$$f' = \sum \sum (c_{ij} - K_i - l_j)x_{ij}$$

$$= f - \sum_i K_i \sum_j x_{ij} - \sum_j l_j \sum_i x_{ij}$$

$$= f - \sum K_i a_i - \sum l_j b_j$$

and the last two terms are constants so f' and f are minimised together.

This has particular application to the special transportation problem known as the assignment problem.

Example 4.6 There are n men to be assigned to n jobs, one man to each job, and a matrix $\{c_{ij}\}$ expresses the unsuitability of man i for job j. We require the assignment which minimises the sum of the unsuitabilities.

We can reduce this to the usual form by putting $x_{ij} = 1$ if man i is assigned to job j, $x_{ij} = 0$ otherwise, and then solving: minimise $\sum \sum c_{ij}x_{ij}$, subject to

$$\left.\begin{array}{ll} \sum_{j=1}^{n} x_{ij} = 1 \quad \text{all } i & \\ & x_{ij} \geqslant 0 \\ \sum_{i=1}^{n} x_{ij} = 1 \quad \text{all } j. & \end{array}\right\} \tag{4.22}$$

The difficulty with solving (4.22) is that there are $2n-1$ basic variables, but clearly only n 1's, and hence every feasible solution is highly degenerate with $n-1$ basic zeros. The method of Section 4.2.2 can still be used but with some wasteful zero changes, that is changes of basis but not of value. It is possible to perturb the problem, and to solve, for example:
minimise $\sum \sum c_{ij}x_{ij}$, subject to

$$\left.\begin{array}{l} \sum_j x_{ij} = 1 + i\varepsilon, \quad \sum_i x_{ij} = 1, \quad j \neq n, \\ \sum_i x_{in} = 1 + \tfrac{1}{2}n(n+1)\varepsilon. \quad x_{ij} \geqslant 0 \end{array}\right\}$$

An alternative method is to make use of the property above. If all c_{ij} are positive, then we can define K_i, l_j so that $c'_{ij} = 0$ in some cells, $c'_{ij} \geqslant 0$ everywhere. In the amended problem, then,

$$f' = \sum\sum c'_{ij} x_{ij} \geqslant 0$$

since $x_{ij} \geqslant 0$. If a feasible set of x_{ij} can be found for which $f' = 0$, this must be the optimal set; and this can be done if $c'_{ij} = 0$ in a set of n independent cells, that is, cells having no row or column in common. It will be seen that K_i, l_j fulfil the same role as the dual variables already used, since $K_i + l_j = c_{ij}$ in a basic cell. The advantage is that we do not need to find all the $2n-1$ basic cells – if we can find a set of n independent basic cells, that is enough to define the solution, which minimises f', and so also minimises f.

4.2.5 Assignment algorithm

The algorithm for this method must then carry out the following sequence:

(1) Produce a set of K_i, l_j which make some $c'_{ij} = 0$, all $c'_{ij} \geqslant 0$.

(2) Test whether there are n independent zeros. If so, the solution is reached. If not, proceed to (3).

(3) Adjust K_i, l_j so as to improve the solution.

In carrying out (3), we make use of Konig's theorem, which states: in a two-dimensional array, the maximum number of independent elements is the same as the minimum number of lines covering all elements.

This can be illustrated using the arrays in Fig. 4.3. In Fig. 4.3(a) the 0's can be covered by 3 lines so there are at most 3 independent elements. In Fig. 4.3(b), there are 4 independent elements – shown boxed.

Proof of Konig's theorem This is a disguised form of the duality theorem, as would be expected from its form, so we give a duality proof; the original proof was a combinatorial one. For the given two-dimensional array, let $x_{ij} = 1$ if there is an element in cell ij and 0 otherwise, and let the set T be the set of i, j, for which $x_{ij} = 1$. Visualise a network (Fig. 4.4) with nodes i, each linked to a common source, S_1, and nodes j, each linked to a common sink, S_2, all these links having capacity 1; and suppose the nodes i, j are connected when $i, j \in T$ and not otherwise.

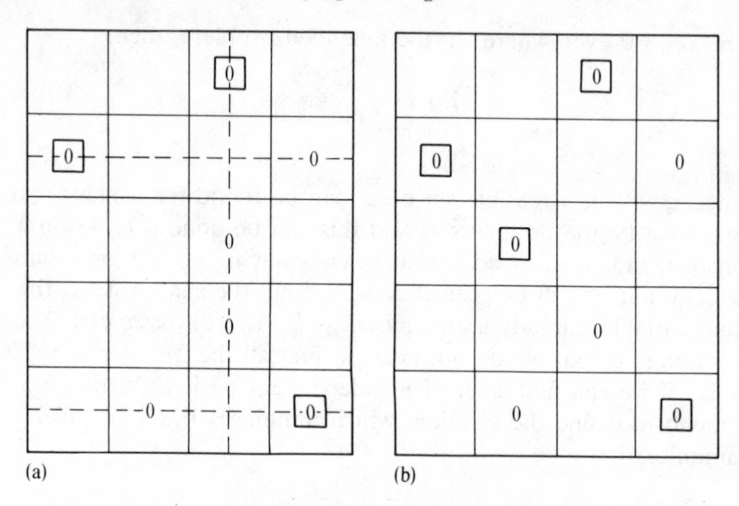

Fig. 4.3. König's theorem.

Consider the problem of finding a flow through this network from the source S_1 to the sink S_2.

If the flow in link ij is t_{ij}, then we must have, for any i, $i = 1, 2, \ldots, n$,

$$\sum_{i,j \in T} t_{ij} \leq 1 \tag{4.23}$$

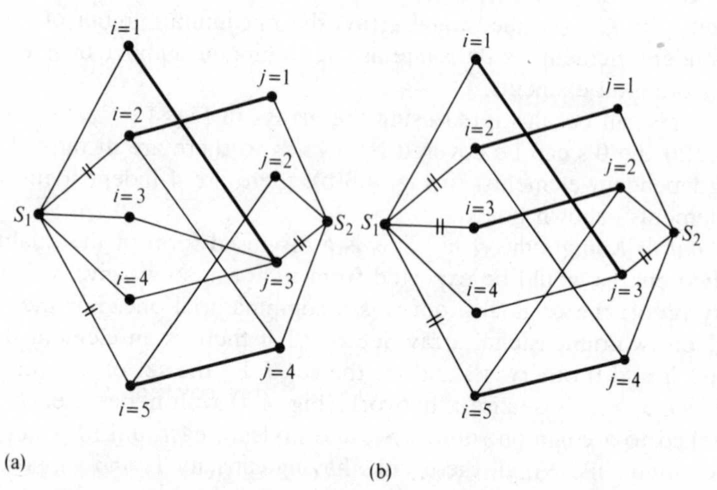

Fig. 4.4. Network corresponding to König's theorem.

For any $j, j=1, 2, \ldots m$,

$$\sum_{i,j \in T} t_{ij} \leqslant 1 \tag{4.24}$$

and for any i, j,

$$0 \leqslant t_{ij} \leqslant x_{ij}.$$

These follow since there is a limit 1 on the material reaching i or leaving j, because of the capacity limits on the links with the source and the sink. Consider then the primal problem,

$$\text{maximise} \quad f = \sum_{i=1}^{n} \sum_{j=1}^{m} t_{ij} \quad \text{subject to conditions, (4.23), (4.24).}$$

The optimal f is then the maximum number of independent elements, since when $t_{ij}=1$, no other t_{ij} with the same i or j can also be 1, because of (4.23), (4.24), and $t_{ij}=1$ only where $x_{ij}=1$.

The dual problem is

$$\begin{aligned} u_i + v_j &\geqslant 1 \quad \text{for} \quad i, j \in T \\ u_i, v_j &\geqslant 0 \end{aligned} \tag{4.25}$$

$$\text{minimise} \quad \sum_{i=1}^{n} u_i + \sum_{j=1}^{m} v_i = g.$$

The function g represents the number of lines covering all (ij), $i, j \in T$, for any such i, j needs either $u_i = 1$ or $v_j = 1$ to satisfy the inequality, (4.25). (If both are 1, the element is covered by 2 lines.) Then since f, g occur in dual problems, it follows from the duality theorem that

$$\max f = \min g;$$

that is,

$$\max \sum \sum t_{ij} = \min \sum u_i + \sum v_j;$$

that is,

maximum number of independent elements

= minimum number of lines covering all elements.

This result can be used to implement the Hungarian method for the assignment problem; the successive steps are now listed, and then a worked example is given.

Step 1 Produce a set of K_i, l_j by choosing $l_j = \min_i c_{ij}$ and $K_i = \min_j (c_{ij} - v_j)$, thus ensuring at least one zero c'_{ij} in each row and column.
Step 2 Find a maximal set of independent zeros of c'_{ij} and star them. If there are n, they define the solution; if there are $m < n$, then all the zeros can be covered by m lines, by Konig's theorem. Initially insert m lines covering the columns containing the m starred zeros. Consider an uncovered column, which must contain at least one zero (by step 1); cover the row containing that zero, and uncover the column containing the corresponding 0*. Repeat until all zeros are covered.
Step 3 Produce a new set of K_i, l_j as follows. Consider values of c'_{ij} uncovered, which by the construction are all positive and have a smallest value h. Increase l_j by h on all uncovered columns and decrease K_i by h on all covered rows. This process retains the starred zeros, produces at least one new zero, and increases the dual objective $\sum K_i + \sum l_j$ by amount

$$h\begin{bmatrix} \text{no. of} & - & \text{no. of} \\ \text{uncovered columns} & & \text{covered rows} \end{bmatrix} = h(n-m). \qquad (4.26)$$

Example 4.7 Solve the assignment problem, with the given matrix, by the Hungarian method.

$$\{c_{ij}\} = \begin{bmatrix} 9 & 22 & 58 & 11 & 19 & 27 \\ 43 & 78 & 72 & 50 & 63 & 48 \\ 41 & 28 & 91 & 37 & 45 & 33 \\ 74 & 42 & 27 & 49 & 39 & 32 \\ 36 & 11 & 57 & 22 & 25 & 18 \\ 3 & 56 & 53 & 31 & 17 & 28 \end{bmatrix}.$$

Step 1 $\mathbf{K}^T = (0, 30, 15, 0, 0, 0)$, $\mathbf{l}^T = (3, 11, 27, 11, 17, 18)$. These give a dual objective value of 132.
Step 2

$$\{c'_{ij}\} = \begin{bmatrix} 6 & 11 & 31 & 0^* & 2 & 9 \\ 10 & 37 & 15 & 9 & 16 & 0^* \\ 23 & 2 & 49 & 11 & 13 & 0 \\ 71 & 31 & 0^* & 38 & 22 & 14 \\ 33 & 0^* & 30 & 11 & 8 & 0 \\ 0 & 45 & 26 & 20 & 0^* & 10 \end{bmatrix}.$$

There are 8 zeros, but only 5 independent zeros, starred, so $m = 5$. The 5 covering lines are shown.
Step 3 $h = 2$. Increase l_j by 2 on $j = 1$, $j = 5$ (uncovered column), decrease K_i by 2 on $i = 6$ (covered). New $\mathbf{K}^T = (0, 30, 15, 0, 0, -2)$, $\mathbf{I}^T = (5, 11, 27, 11, 19, 18)$. New objective 134.

Step 2

$$\{c'_{ij}\} = \begin{bmatrix} 4 & 11 & 31 & 0^* & 0 & 9 \\ 8 & 37 & 15 & 9 & 14 & 0^* \\ 21 & 2 & 49 & 11 & 11 & 0 \\ 69 & 31 & 0^* & 38 & 20 & 14 \\ 31 & 0^* & 30 & 11 & 6 & 0 \\ 0 & 47 & 28 & 22 & 0^* & 12 \end{bmatrix}.$$

There are now 9 zeros, including all the previously starred zeros, but still $m = 5$. The 5 covering lines are now as shown.
Step 3 $h = 6$. New $\mathbf{K}^T = (-6, 30, 15, 0, 0, -8)$, $\mathbf{I}^T = (11, 11, 27, 17, 25, 18)$. New objective 140.
Step 2

$$\{c'_{ij}\} = \begin{bmatrix} 4 & 17 & 37 & 0^* & 0 & 15 \\ 2 & 37 & 15 & 3 & 8 & 0^* \\ 15 & 2 & 49 & 5 & 5 & 0 \\ 63 & 31 & 0^* & 32 & 14 & 14 \\ 25 & 0^* & 30 & 5 & 0 & 0 \\ 0 & 53 & 34 & 22 & 0^* & 18 \end{bmatrix}.$$

There are now 10 zeros, but again $m = 5$.
Step 3 $h = 2$. New $\mathbf{K}^T = (-8, 30, 15, 0, -2, -10)$, $\mathbf{I}^T = (13, 13, 27, 19, 27, 18)$. New objective 142.
Step 2

$$\{c'_{ij}\} = \begin{bmatrix} 4 & 17 & 39 & 0^* & 0 & 17 \\ 0^* & 35 & 15 & 1 & 6 & 0 \\ 13 & 0 & 49 & 3 & 3 & 0^* \\ 61 & 29 & 0^* & 30 & 12 & 14 \\ 25 & 0^* & 32 & 5 & 0 & 2 \\ 0 & 53 & 36 & 22 & 0^* & 20 \end{bmatrix}.$$

The 6 independent zeros are shown starred, and the solution is attained. (Actually we have the whole solution since there are 11 zeros altogether.) The final solution has a primal objective $11+43+33+27+11+17=142$, the same as the dual objective.

4.3 Game theory

A "game" is defined to be a contest between opponents in which each has a number (finite or infinite) of courses of action, "strategies", and the outcome of any combination of strategies is known beforehand. The problem is to determine the best strategy for each player given that they have, in general, competing objectives and that each plays independently.

We specialise here to a two-person game – n-person games involve consideration of coalitions and other mutual arrangements. Assume also that the game is zero-sum, such that the benefit to one player is always equal to the loss to the other player. This is by no means always so in a practical situation: there is a classic example – the prisoner's dilemma – which can be stated as follows:

Two prisoners, suspected of having jointly committed a crime, are interrogated separately. Each knows that they will both be released if neither confesses. If one confesses and the other does not, then the one who confesses will get a light penalty and the other a heavy penalty; if both confess, each receives a light penalty. The table of consequences is then as shown in Table 4.4.

In this case, the best result for both follows if both choose not to confess. However, psychologically it is likely that each will in fact confess for fear of being let down by the other and so incurring a heavy penalty.

For two-person zero-sum games, however, this sort of anomaly cannot happen, and we can prove the existence of an optimal

Table 4.4

Prisoner A \ Prisoner B	Confess	Not confess
Confess	A light B light	A light B heavy
Not confess	A heavy B light	A release B release

strategy for each player, and a result which neither player can be certain of improving. A single function of the two strategies, giving the benefit to one player, defines the game. We shall deal in what follows only with a finite set of strategies, so that the effects of any combination of strategies can be exhibited as a matrix, the "payoff matrix" **A**.

4.3.1 Definition of problem

Suppose P_1 has m strategies, P_2 has n, and **A**, an $m \times n$ matrix, gives the gain to P_1. Then

$$\begin{aligned}(\text{gain to } P_1 \text{ playing } i) &= a_{ij} \quad \text{when } P_2 \text{ plays } j \\ &\geqslant \min_j a_{ij}.\end{aligned} \tag{4.27}$$

Similarly whatever P_1 does

$$(\text{loss to } P_2 \text{ playing } j) \leqslant \max_i a_{ij}. \tag{4.28}$$

For any **A** we have found (3.2) that

$$\min_j \max_i a_{ij} \geqslant \max_i \min_j a_{ij}.$$

So suppose first that this is an equality, that is there exists a such that

$$\min_j \max_i a_{ij} = a = \max_i \min_j a_{ij}, \tag{4.29}$$

where

$$a = a_{i_0 j_0}.$$

Such a value is called a saddle point of the matrix. In this case (4.27), (4.28) mean

$$\begin{aligned} a_{i_0 j} = \text{gain to } P_1 \text{ playing } i_0 &\geqslant \min_j a_{i_0 j} \\ &= a \\ &\geqslant \max_i a_{i j_0} \\ &\geqslant \text{loss to } P_2 \text{ playing } j_0, \end{aligned} \tag{4.30}$$

and so P_1 playing i_0 must win at least a, P_2 playing j_0 will lose not more than a, and the result is a if both play their optimum strategies, i_0, j_0.

Table 4.5

P_1's strategies \ P_2's strategies	1	2	3	$\min_j a_{ij}$
1	5	1	−1	−1
2	2	4	−2	−2
3	1	−1	3	−1
$\max_i a_{ij}$	5	4	3	

However in general a saddle point does not exist, that is

$$\min_j \max_i a_{ij} > \max_i \min_j a_{ij} \tag{4.31}$$

The situation is now different: in playing pure strategy i_0 which maximises $\min_j a_{ij}$, P_1 can guarantee to win at least $\max_i \min_j a_{ij}$, and similarly P_2 can guarantee to lose not more than $\min_j \max_i a_{ij}$, but since the two values are unequal this cannot define an equilibrium way of playing. Consider a simple example.

Example 4.8 Find the best way of playing the game whose payoff matrix $\mathbf{A}$ is as given in Table 4.5, each player having 3 strategies.

P_1 playing strategy 1 (or 3) gains at least −1, P_2 playing strategy 3 loses at most 3, but clearly neither of these methods provides the optimum result for both players.

We therefore define a "mixed strategy" $\mathbf{x}$ in which a player plays his i-th pure strategy with relative frequency x_i; then $x_i \geq 0$, $\sum x_i = 1$. Assume that he also imposes some random order, so that his opponent cannot predict which strategy he will adopt at the next play – for instance, a mixed strategy $\mathbf{x}^T = (1/6, 1/2, 1/3)$ can be implemented by throwing a dice and playing strategy 1 if the result is 1, 2 if the result is 2, 3 or 4, and 3 if the result is 5 or 6. The result of P_1 playing a mixed strategy $\mathbf{y}$ and P_2 playing $\mathbf{x}$ is then a gain to P_1 of $\mathbf{y}^T\mathbf{A}\mathbf{x}$ in the long run. It is the fundamental result of game theory that there is always an optimal pair of mixed strategies, $\mathbf{x}^*$, $\mathbf{y}^*$ and a value, v^*, for the game, such that

$$v^* = \mathbf{y}^{*T}\mathbf{A}\mathbf{x}^*. \tag{4.32}$$

We prove this by use of the duality theorem.

4.3.2 Determination of optimal strategies

Note first that if every element of $\mathbf{A}$ is increased by a constant, c, to give a new matrix $\mathbf{A}'$, $a'_{ij} = a_{ij} + c$, then

$$\mathbf{y}^{\mathrm{T}}\mathbf{A}'\mathbf{x} = \mathbf{y}^{\mathrm{T}}\mathbf{A}\,\mathbf{x} + c\sum_i \sum_j y_i x_j = \mathbf{y}^{\mathrm{T}}\mathbf{A}\,\mathbf{x} + c \tag{4.33}$$

since $\sum x_j = 1$, $\sum_i y_i = 1$. Hence the effects of $\mathbf{y}$ and $\mathbf{x}$ on the games defined by $\mathbf{A}$ and $\mathbf{A}'$ are the same; and for any $\mathbf{A}$, by choosing c large enough, we can produce an $\mathbf{A}'$ which will give a positive gain to P_1. We therefore consider only games for which this is so.

Now specifying the problem for each player we have for P_1: find $\mathbf{y}$ such that $\mathbf{y} \geqslant 0$, $\mathbf{1}_m^{\mathrm{T}}\mathbf{y} = 1$, and v, a maximum, for which

$$\mathbf{y}^{\mathrm{T}}\mathbf{A} \geqslant v\mathbf{1}_n. \tag{4.34}$$

Here $\mathbf{1}_r$ is the r-vector consisting entirely of $\mathbf{1}$'s.

Similarly for P_2: find $\mathbf{x}$ such that $\mathbf{x} \geqslant 0$, $\mathbf{1}_n^{\mathrm{T}}\mathbf{x} = 1$, and v', a minimum, for which

$$\mathbf{A}\mathbf{x} \leqslant v'\mathbf{1}_m. \tag{4.35}$$

As above, we can assume $v, v' > 0$ and so, writing $y'_i = y_i/v$, $x'_j = x_j/v'$, these become

$$\begin{aligned} &\mathbf{y}'^{\mathrm{T}}\mathbf{A} \geqslant \mathbf{1}_n,\ \mathbf{y}' \geqslant 0,\ \mathbf{1}_m^{\mathrm{T}}\mathbf{y}' = 1/v \quad \text{to be minimised} \\ &\mathbf{A}\mathbf{x}' \leqslant \mathbf{1}_m,\ \mathbf{x}' \geqslant 0,\ \mathbf{1}_n^{\mathrm{T}}\mathbf{x}' = 1/v' \quad \text{to be maximised.} \end{aligned} \tag{4.36}$$

These now form a pair of dual problems and so, by the duality theorem, the optimal values of v and v' are the same, v^*, and there is an optimal strategy for each player, $\mathbf{x}^*$, $\mathbf{y}^*$, such that

$$\mathbf{y}^{*\mathrm{T}}\mathbf{A}\mathbf{x} \geqslant \mathbf{y}^{*\mathrm{T}}\mathbf{A}\mathbf{x}^* \geqslant \mathbf{y}^{\mathrm{T}}\mathbf{A}\mathbf{x}^* \tag{4.37}$$

and

$$\mathbf{y}^{*\mathrm{T}}\mathbf{A}\mathbf{x}^* = v^*.$$

Further, the simplex method applied to (4.36) will produce both sets of strategies. Note that each problem clearly has a feasible solution and hence an optimal solution always exists.

Solution to Example 4.8 By inspection of the matrix $\mathbf{A}$ in Example 4.8, we can certainly ensure a positive gain by putting $c = 2$ so that

$$\mathbf{A}' = \begin{bmatrix} 7 & 3 & 1 \\ 4 & 6 & 0 \\ 3 & 1 & 5 \end{bmatrix}.$$

Table 4.6

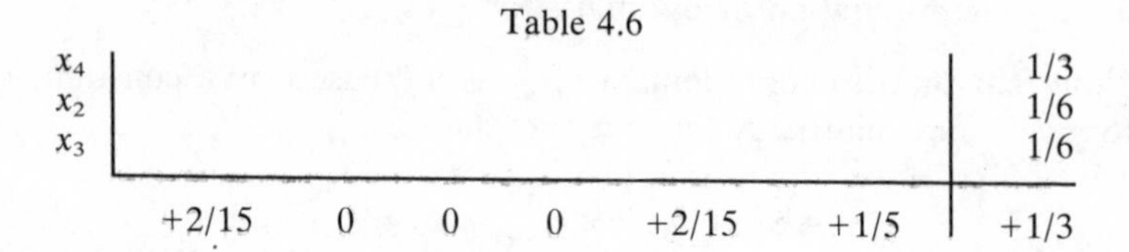

x_4							1/3
x_2							1/6
x_3							1/6
	+2/15	0	0	0	+2/15	+1/5	+1/3

The final simplex tableau for the problem

$$\mathbf{Ax}' \leqslant \mathbf{1}, \quad \mathbf{x}' \geqslant 0, \quad \max + \mathbf{1}^{\mathrm{T}}\mathbf{x}'$$

is found to be as shown in Table 4.6. Hence $1/v'^*$ for this game is 1/3, $v'^* = 3$

$$x_1'^* = 0, \quad x_2'^* = 1/6, \quad x_3'^* = 1/6,$$

and so

$$x_1^* = 0, \quad x_2^* = 1/2, \quad x_3^* = 1/2.$$

Similarly, reading off the dual variables, $\mathbf{y}^{*\mathrm{T}} = (0, 2/5, 3/5)$. Finally, for the original game, $v^* = v'^* - 2 = 1$, so the optimum result is a gain of 1 to P_1 at each play.

4.3.3 Further aspects of game theory

Game theory has been used directly in studies of warfare and of industrial competition. It also forms a part of decision theory; if a decision maker is faced with external conditions over which he has no control (for instance a farmer planning future crops) then one form of decision rule is to regard himself as playing a game against "nature" and adopt the minimax strategy above, so as to guarantee a "least worst" result. This assumes knowledge of the range of possible external conditions, and of the outcome of each of his strategies in conjunction with these. Since nature is not in fact a hostile opponent, there may be more useful forms of decision rule; however we shall not consider this topic further here.

Another aspect of game theory which should be mentioned is dependence on chance. Restricting discussion simply to conventional games, we can distinguish between those with perfect information, e.g. noughts and crosses, draughts, chess, go, and games with a random element, e.g. poker, baccarat, bridge. It can be proved that games in the first category always have a saddle point, that is there is an optimal pure strategy for each player. The enumeration of the pure strategies is an impossibly large undertaking in draughts or chess, since it must include the reaction to any move of the opponents; you might try enumerating the strategies for noughts and crosses, but this is still quite a substantial operation. A simplifying consideration is the idea of dominated strategies; if the elements in one row of **A** are all less than the corresponding elements in another row, that is if

$$a_{i_1 j} < a_{i_2 j}, \quad \text{all } j,$$

then P_1 will never play strategy i_1, and it need not be considered. Similarly if

$$a_{ij_1} > a_{ij_2}, \quad \text{all } i,$$

we can delete strategy j_1 for P_2. If $<$, $>$ are replaced by $\leqslant$, $\geqslant$ the dominated strategy cannot affect the value of the game.

If we now consider games of the second kind, the payoff matrix must take into account the external chance effects as well as the strategies. Consider the following game.

Example 4.9 P_1 draws at random one of two cards, Hi or Lo. He can either pass or bet. If he passes, he must show his card, and wins a with Hi, loses a with Lo. If he bets, P_2 can either pass or call. If P_2 passes, P_1 wins a. If P_2 calls, P_1 wins b with Hi, loses b with Lo, $b > a$.

The strategies available to P_1 are

S_1: pass with Lo, pass with Hi

S_2: pass with Lo, bet with Hi

S_3: bet with Lo, pass with Hi

S_4: bet with Lo, bet with Hi.

The strategies available to P_2 are

T_1: pass if P_1 bets

T_2: call if P_1 bets

The payoff is as shown in Table 4.7

Table 4.7

	T_1	T_2
S_1:	$\frac{1}{2}(-a)+\frac{1}{2}(a)=0$	0
S_2:	$\frac{1}{2}(-a)+\frac{1}{2}(a)=0$	$\frac{1}{2}(-a)+\frac{1}{2}(b)=\frac{1}{2}(b-a)$
S_3:	$\frac{1}{2}(a)+\frac{1}{2}(a)=a$	$\frac{1}{2}(-b)+\frac{1}{2}(a)=-\frac{1}{2}(b-a)$
S_4:	$\frac{1}{2}(a)+\frac{1}{2}(a)=a$	$\frac{1}{2}(-b)+\frac{1}{2}(b)=0$

where for example the entry for S_2, T_2 is made up of

P_1 gets Lo and passes, losing a – probability 1/2

and

P_1 gets Hi and bets, P_2 calls, P_1 wins b – probability 1/2,

producing an expected gain to P_1 of $\frac{1}{2}(b-a)$.

You can verify that strategy S_1 is dominated by S_2, and S_3 by S_4, and that the optimal way of playing is

P_1 plays S_2 with frequency $2a/(a+b)$,
S_4 with frequency $(b-a)/(a+b)$

P_2 plays T_1 with frequency $(b-a)/(a+b)$,
T_2 with frequency $2a/(a+b)$.

The value is $a(b-a)/(a+b)$, and P_1 can expect to win this per game over a long sequence of games.

4.4 Separable programming

A problem with a nonlinear objective and constraints can sometimes be dealt with by using piecewise linear approximations; the number of variables is increased but linear programming methods can be used to find an approximate solution.

Given a set of reference values (knots) of x,

$$x_1 < x_2 < \cdots < x_{k-1} < x_k,$$

any x in the range $x_1 \leqslant x \leqslant x_k$ can be written as a linear combination

$$x = \sum w_i(x) x_i,$$

where $w_i(x) \geqslant 0, \sum w_i(x) = 1$, and at most *two* of the $w_i(x)$ are nonzero for any given x. The nonzero w_i must be for adjacent i; for $x_i \leqslant x \leqslant x_{i+1}$, we have

$$w_i(x) = (x_{i+1} - x)/(x_{i+1} - x_i), \quad w_{i+1}(x) = (x - x_i)/(x_{i+1} - x_i)$$

and all other w's are zero. Any function $f(x)$ can then be approximated in terms of its values $f_1, f_2, \ldots, f_k$ at the knots. The expression $z(x) = \sum w_i f_i$ is the first-order Lagrange interpolation formula for f, with an error, for $x_i \leqslant x \leqslant x_{i+1}$ of

$$\tfrac{1}{2}(x - x_i)(x - x_{i+1}) f''(x'),$$

where x' is some value in the range $x_i \leqslant x' \leqslant x_{i+1}$ (see Note 4.1). Since the linearity of the approximation is essential, the error can be reduced only by reducing the spacing, and so increasing the number, of the knots.

This can be applied to a nonlinear programming problem in $\mathbf{x} \in R^n$, for which the constraints $\mathbf{c}$ and the objective f are *separable* – that is, $c_i(\mathbf{x}) = \sum c_{ij}(x_j)$, $f(\mathbf{x}) = \sum f_j(x_j)$, in terms of the components x_j of the vector $\mathbf{x}$. Such a problem can be made into a linear problem in the w_{ji}, the coefficients associated with the i-th knot for x_j. This can be solved by the ordinary simplex method, subject only to the restriction that the w_{ij} can, for any i, only be nonzero for two adjacent values of j. The w-solution can then be transformed back to give approximate values of $\mathbf{x}^*$, f^*.

Example 4.10 Minimise

$$f(x) = (x_1 - 2)^2 + 4(x_2 - 1)^2$$

subject to

$$2x_1 + (x_2 + 1)^2 \leqslant 4, \quad x_1, x_2 \geqslant 0.$$

Since the ranges for x_1, x_2 are 0, 1.5 and 0, 1 respectively, take knots for x_1 at 0, 0.5, 1, 1.5 and for x_2 at 0, 0.5, 1. Then the values of the nonlinear functions $f_1(x) = (x-2)^2$, $f_2(x) = 4(x-1)^2$ and $c(x) = (x+1)^2$ can be tabulated at the appropriate knots (Table 4.8).

Table 4.8

x	$f_1(x)$	$f_2(x)$	$c(x)$
0	4	4	1
0.5	2.25	1	2.25
1	1	0	4
1.5	0.25		

The variables x_1, x_2 can be written

$$x_1 = 0.w_{11} + 0.5w_{12} + 1w_{13} + 1.5w_{14}$$

$$x_2 = 0.w_{21} + 0.5w_{22} + 1w_{23},$$

and the linear approximations to the functions are

$$f_1(x) \simeq p_1(x) = 4w_{11} + 2.25w_{12} + 1w_{13} + 0.25w_{14}$$

$$f_2(x) \simeq p_2(x) = 4w_{21} + 1w_{22} + 0.w_{23}$$

$$c(x) \simeq q(x) = 1w_{21} + 2.25w_{22} + 4w_{23}.$$

Table 4.9

T$_1$:

	w_{11}	w_{12}	w_{13}	w_{14}	w_{21}	w_{22}	w_{23}	w_0	
w_0	0	1	2	3	0	1.25	3	1	3
w_{11}	1	1	1	1	0	0	0	0	1
w_{21}	0	0	0	0	1	①	1	0	1
	0	1.75	3	3.75	0	3	4	0	8

The approximate problem is then to minimise

$$z = 4w_{11} + 2.25w_{12} + w_{13} + 0.25w_{14} + 4w_{21} + w_{22}$$

subject to

$$\begin{aligned} w_{12} + 2w_{13} + 3w_{14} + w_{21} + 2.25w_{22} + 4w_{23} &\leqslant 4 \\ w_{11} + w_{12} + w_{13} + w_{14} &= 1 \\ w_{21} + w_{22} + w_{23} &= 1 \end{aligned}$$

with $w \geqslant 0$ and the additional restriction that only two adjacent w_{1j}, w_{2j} can be nonzero.

The initial simplex tableau (Table 4.9) is formed by including a slack variable w_0 in the first inequality and taking w_0, w_{11} and w_{21} as basic variables. The variable w_{23} cannot enter the basis since w_{21} is already basic, and neither can w_{14}; so we introduce w_{22}, which is allowable. The pivot is shown, and the next tableau is produced in the usual way. Continuing the process gives a final tableau **T*** (Table 4.10), which gives

$$w_{12} = 0.25, \quad w_{13} = 0.75, \quad w_{22} = 1, \quad z^* = 2.3125$$

and, substituting back,

$$x_1^* = 0.875, \quad x_2^* = 0.5.$$

The exact solution to this problem, for comparison, is

$$x_1^* = 0.8, \quad x_2^* = 0.55, \quad \text{and} \quad f = 2.25.$$

It can be shown (see Problem 20), that if f is convex and **c** is

Table 4.10

T*:

w_{12}	2	1	0	−1	1.25	0	−1.75	−1	0.25
w_{13}	−1	0	1	2	−1.25	0	1.75	1	0.75
w_{22}	0	0	0	0	1	1	1	0	1
	−0.5	0	0	−0.5	−1.4375	0	−1.1875	−1.25	2.3125

concave, then the linear approximate solution has a value which is not less than the exact minimum, that is that $z^* \geq f^*$.

Note

4.1 The error in Lagrange first-order interpolation

This is proved by considering, for fixed x in the range $x_i \leq x \leq x_{i+1}$, the function

$$F(t) = f(t) - z(t) - \{f(x) - z(x)\}W(t)/W(x)$$

where

$$W(t) = (t - x_i)(t - x_{i+1}).$$

Since $F(x_i) = F(x_{i+1}) = F(x) = 0$, $F''(t) = 0$ at least for one value $t = x'$, where $x_i \leq x' \leq x_{i+1}$. Since $z'' = 0$ for all t, $W'' = 2$ for all t, this gives

$$f(x) - z(x) = \tfrac{1}{2}W(x)f''(x')$$

Problems

(1) In the resource allocation problem, what is the economic significance of the fact that the dual variable corresponding to an equality constraint is unrestricted in sign?

(2) A manufacturer uses two materials A, B and makes two types of the same product, P_1, P_2. Each type can be processed through either of two machines M_1, M_2. How should production be scheduled so as to maximise return, given the requirements and costs listed in Table 4.11?

Table 4.11

	Product P	Product P	Total available (tonnes)	Cost
Material A	20%	40%	100	£6/tonne
B	80%	60%	500	£10/tonne
on machine M_1	1 h/tonne	2 h/tonne	200 h	£4/h
M_2	3 h/tonne	4 h/tonne	300 h	£3/h
selling at	£20/tonne	£22/tonne		

At what prices would it pay to buy extra material or to hire extra machine time?

(3) Paper is produced in a continuous strip of width W. It is required to cut this so as to supply lengths l_1 of width $w_1, \dots, l_n$ of width w_n where $w_j < W$, all j. Write as an LP problem the determination of the most economical way to do this, i.e. the way in which the consequent area of waste is minimised.

(4) A set of observations $(x_i, a_i, b_i, \dots, t_i)$ is given, $i = 1, 2, \dots, n$. It is required to determine constants $\alpha, \beta, \dots, \tau$ such that

$$\sum_{i=1}^{n} |x_i - \alpha a_i - \beta b_i - \cdots - \tau t_i|$$

is minimised. Show that this is equivalent to an LP problem.

(*Hint*: Consider $x_i - \alpha a_i - \beta b_i - \cdots - \tau t_i = P_i - Q_i$, $P_i, Q_i > 0$)

(5) A factory has the choice of 3 different systems for making a particular product.

(1) A worker runs 3 machines of type A, each costs £20/week to run, each produces 100 units/week, the worker is paid £60/week.

(2) A worker runs 5 machines of type A, each costs £20/week, each produces 70 units/week, the worker is paid £70/week.

(3) A worker runs 2 machines of type B, each costs £35/week, each produces 150 units/week, the worker is paid £75/week. There are 45 machines of type A, 10 of type B, 15 workmen. How should they be used if the objective is to maximise output? What is the significance of the dual variables in this case? Every unit produced can be sold at £1; is the scheduling different if the objective is to maximise profit? What then is the significance of the dual variables?

(6) A closed economy may be considered as made up of a number of firms, F_i, $i = 1, \dots, n$, each producing a different product P_i. Some production from each firm is required to satisfy the needs of other firms: to produce x_i units of P_i requires $a_{ji}\, x_i$ units of P_j, $j = 1, \dots, n$. The rest of the production is consumed and at least b_i units of P_i are required for that, $b_i \geqslant 0$. Labour requirements are $a_{0i}\, x_i$ to produce x_i units of P_i.

(a) Write as an LP problem the determination of values of x_i which will satisfy the demands b_i, and the requirements of all the firms, and will minimise total labour required.
(b) State the conditions for this LP to have a finite optimum and show that then all x_i must be basic.
(c) Write the optimal solution in matrix form.
(d) Write down the dual of this problem and show that the dual variables are independent of $\mathbf{b}$.

(7) Three sources S_1, S_2, S_3 produce quantities 13, 9, 12 of material and distribute this to four destinations D_1, D_2, D_3, D_4 with demands 5, 10, 9, 10 and a cost matrix as shown in Table 4.12.

Table 4.12

	D_1	D_2	D_3	D_4
$\{c_{ij}\} = S_1$	3	4	8	5
S_2	1	7	10	7
S_3	3	7	11	9

What is the best distribution? What modification is necessary if S_2 produces 11 units?

(8) There are three sources S_1, S_2, S_3 producing respectively 20, 20 and 200 units, and four destinations D_1, D_2, D_3, D_4 requiring respectively 200, 10, 60 and 20 units. Passage between S_1 and D_3, S_2 and D_1, and S_2 and D_2 is impossible: for other routes the cost matrix is as given in Table 4.13.

Table 4.13

	D_1	D_2	D_3	D_4
$\{c_{ij}\} = S_1$	30	5		40
S_2			5	90
S_3	20	40	10	10

Find the minimum cost solution, and also the unsatisfied demand at each destination.

(9) (The caterer problem.) A caterer is faced with the problem of supplying clean tablecloths for the next N days. He knows the number n_j needed on day j, $j = 1$ to N; and can choose between

the alternatives

(1) buy fresh at a cost c per unit,

(2) have laundered with a fast 24 h service at cost of l_f per unit,

(3) have laundered with a slow, 48 h service at cost l_s per unit where $c > l_f > l_s$. Formulate the problem of satisfying demand at minimum cost as a transportation problem. (*Hint*: Take sources as being cloths new, or available but dirty, at the end of each of the N days: destinations as requirements for each of the N days, and a sink for dirty cloths. Assume he has no cloths to begin with and that all he buys are dirty at the end. Note that some cells of the table are impossible and so must carry infinite cost.)

(10) A transportation problem with given cost matrix c_{ij} and source and destination totals a_i, b_j such that $\sum a_i = \sum b_j$, has a minimum cost solution x_{ij}^*, where $C^* = \sum\sum x_{ij}^* c_{ij}$

(a) Transitions between the sources and also between the destinations are now allowed, with cost coefficients $c_{i_1 i_2}$ and $c_{j_1 j_2}$. Under what conditions will this reduce the total cost C^*?

(b) Alternatively, additional depots p_k are allowed, with cost coefficients c_{ik}, c_{kj}. Under what conditions will this reduce C^*?

(c) Starting with the solution of Problem 7, test your answers to (a) and (b) by modifying the problem and solving the modified one.

(11) (Equipment replacement.) The cost of renting and maintaining an item of equipment from the beginning of year i to the beginning of year j is c_{ij}. The equipment is needed over the next N years. In which years should a new model be rented to make the overall cost lower? Show that this problem is mathematically equivalent to the problem of finding the shortest route through a network, and that this is a transshipment problem.

(12) Consider the problem of assigning 5 operators to 5 machines. The assignment costs are given by Table 4.14. The

Table 4.14

Operator/machine	1	2	3	4	5
1	5	5	—	2	6
2	7	4	2	3	4
3	9	3	5	—	3
4	7	2	6	7	2
5	6	5	7	9	1

blanks represent impossible assignments. Solve the problem of minimum cost (a) by the Hungarian method; (b) by a perturbed procedure.

(13) Find the optimal strategies and the value of the two games whose payoff matrices are given (a) in Table 4.15 and (b) in Table 4.16.

Table 4.15

P_1 \ P_2	T_1	T_2	T_3	T_4
S_1	8	6	2	8
S_2	8	9	4	5
S_3	7	5	3	5

Table 4.16

P_1 \ P_2	T_1	T_2	T_3	T_4
S_1	1	9	6	0
S_2	2	3	8	4
S_3	−5	−2	10	−3
S_4	7	4	−2	−5

(14) Colonel Blue and Colonel Red have respectively 2 and 3 units to post in either or both of two locations A and B. If Blue's forces, p, outnumber Red's, q, at one location, the gain to Blue is $(q+1)$, and vice versa. If the forces are equal at one location, the payoff there is zero to each. The total gain is the sum of the gains at the two locations. Units are posted whole so that p, q are integers. What is the best strategy for each Colonel?

(15) An $m \times m$ matrix is called a latin square if each row and column contains each of the integers from 1 to m. Show that a game which has this as its payoff matrix has the value $\frac{1}{2}(1+m)$.

(16) Show that a game which has a skew symmetric payoff matrix must have $v^*=0$, and that if $(\mathbf{x}^*, \mathbf{y}^*)$ are optimal, so are $(\mathbf{y}^*, \mathbf{x}^*)$.

(17) Two players fight a duel: they face each other $2n$ paces apart and each has a single bullet in his gun. At a signal each may fire. If either is hit or if both fire the game ends. Otherwise both advance one pace and may again fire. The game of course ends anyway by the time n paces have been taken. The probability of either hitting his opponent if he fires after the i-th pace

forward is i/n. The payoff is $+1$ to a player who survives after his opponent is hit, and 0 if neither or both are hit. The guns are silent so that neither knows whether or not his opponent has fired.

Show that, if $n=4$, the strategy "shoot after taking two steps" is optimal for both, but that if $n=5$ a mixed strategy is optimal, and find this.

(18) A series of runs is made in each of which a valuable object V is concealed in one of two identical bomber aircraft, P and F. The aircraft fly with F above and protecting P. The enemy have two fighters, each of which can make just one attack on one bomber. If the first fighter attacks P, its chance of destroying P is γ, if it attacks F its chance of destroying F is β ($\beta>\gamma$). If the first attack fails these probabilities are unchanged for the attack by the second fighter; but if the first fighter destroys its target then the probability that the second fighter will destroy the survivor is α ($\alpha>\beta$). The gain to the fighters is the proportion of runs in which V is destroyed.

Show that the fighters have four possible strategies and the bombers two, and write down the payoff matrix.

Given that $\alpha=2/3$, $\beta=1/2$, $\gamma=1/6$, prove that the fighters have only a 5/12 chance of destroying V if the bombers use their optimal strategy, and specify the optimal strategies for both.

(19) (A simplified version of baccarat.) A pack of cards is made up of 1's and 6's in equal proportions. The banker deals 1 card face down to the player, and gives himself a 6. The player after looking at his hand can stand (take no card) or draw (ask for an extra card which is dealt to him face upwards). The banker can then either stand or draw. After this the hands are compared, and the higher value (modulo 10) wins – if the values are equal the result is a tie. Thus if the player has $6+6$ ($=2$) and the banker did not draw, the banker's 6 wins. The player has four strategies – stand with 1 or 6, stand with 1 and draw with 6, draw with 1 and stand with 6, draw with 1 or 6. Each of the banker's strategies must prescribe what he is to do under each of the conditions, player stands, player draws 1, player draws 6. Show that the banker has $2^3=8$ strategies and evaluate the payoff matrix, the elements being (probability of winning – probability of losing). Show that many of the banker's strategies are dominated and find the optimum strategy for each player and the value.

(20) Given that the separable objective function $f(\mathbf{x}) = \sum_j f_j(x_j)$ is convex, and $z(\mathbf{x})$ is the approximating linear function $\sum_j \sum_i w_{ji}(x_j) f_{ji}$, show that, for any $\mathbf{x}$, $z(\mathbf{x}) \geqslant f(\mathbf{x})$. Hence show that if the $c_i(\mathbf{x})$ are concave and separable, then the solution to the problem, minimise $f(\mathbf{x})$ subject to $c(\mathbf{x}) \geqslant 0$, and the solution to the linear approximate problem, minimise $z(\mathbf{x})$ subject to the linear version of the constraints, are related by min $z \geqslant$ min f.

(21) Make a transformation of variables to convert the following problem into a separable form, and determine an approximate solution.

Minimise $\quad f(\mathbf{x}) = 25e^{-x_1} - x_2$

subject to $\quad x_1^2 x_2 \leqslant 16, \quad x_1, x_2 \geqslant 0.$

5

Constrained optimisation

This chapter deals with methods for finding the minimum of a function $f(\mathbf{x})$ subject to constraints $\mathbf{c}(\mathbf{x}) \geqslant 0$. Previous chapters have discussed the special cases of no constraints, and of f and $\mathbf{c}$ both linear. In Section 5.1 we recall the general properties of a solution from Chapter 1, and then describe the two main types of method used for the general problem.

The first, projection methods, are particularly suitable when $f(\mathbf{x})$ is nonlinear but each $c_i(\mathbf{x})$ is linear, and their use under these conditions is described in Section 5.2. An important special case is when $f(\mathbf{x})$ is quadratic and $c_i(\mathbf{x})$ linear, the case known as quadratic programming. Efficient algorithms are available for this and are described in Section 5.3. In Section 5.4 the use of projection methods with nonlinear $c_i(\mathbf{x})$ is briefly discussed.

This general case is often better treated by the second type of method, replacing the constrained problem by a sequence of unconstrained problems. One idea is to modify the objective function $f(\mathbf{x})$ by adding terms which penalise violation of the constraints, and Section 5.5 describes the use of these penalty function methods in sequential unconstrained minimisation. Another way, discussed in Section 5.6, is to use this same idea on the Lagrangian $L(\mathbf{x}, \boldsymbol{\lambda})$ and so produce a function which has the desirable features of $L(\mathbf{x}, \boldsymbol{\lambda})$ but with the great advantage, which the Lagrangian does not enjoy, of having an unconstrained local minimum.

The subject of constrained optimisation is a large and continually developing one, and this chapter can only give a brief treatment of some of the main methods of each type, omitting details of proofs and of numerical implementation. The problems at the end of the chapter contain some of the many other ideas which have been used.

The methods described here all make use of the first, and in some cases the second, partial derivatives of $f(\mathbf{x})$ and $\mathbf{c}(\mathbf{x})$, and so assume that these are readily available. In many practical situations this is not the case: the functions may be lengthy to compute and explicit differentiation may be very difficult. It is necessary then either to use finite difference approximations or direct search methods. No discussion of either of these alternatives is given here, but reference may be made to Swann (1974) for further information.

5.1 General properties of the solution

In Chapter 1 conditions for $\mathbf{x}^*$ to be a constrained minimum for the standard problem:

$$\text{minimise } f(\mathbf{x}) \text{ subject to } \mathbf{c}(\mathbf{x}) \geqslant 0, \mathbf{x} \in R^n, \mathbf{c} \in R^m$$

were found, and these are recalled here.

(a) The active set at $\mathbf{x}^*$ is I, so that

$$I = \{i \mid c_i(\mathbf{x}^*) = 0\}. \tag{5.1}$$

If I is empty then $\mathbf{x}^*$ is an unconstrained minimum. In this case necessary and sufficient conditions are

$$c_i(\mathbf{x}^*) \geqslant 0, \mathbf{g}(\mathbf{x}^*) = 0, \mathbf{G}(\mathbf{x}^*) \text{ positive definite.}$$

(b) In general necessary conditions are that $\mathbf{x}^*$ and some $\boldsymbol{\lambda}^*$ satisfy

$$\mathbf{g}(\mathbf{x}^*) = \sum_{i=1}^{m} \lambda_i^* \mathbf{a}_i(\mathbf{x}^*) \tag{5.2}$$

with

$$\mathbf{c}(\mathbf{x}^*) \geqslant 0, \lambda_i^* \geqslant 0, \lambda_i^* c_i(x^*) = 0. \tag{5.3}$$

An equivalent statement to (5.3) is

$$c_i(x^*) > 0, \lambda_i^* = 0, i \notin I; \lambda_i^* \geqslant 0, i \in I. \tag{5.4}$$

Note that the condition (5.2) is appropriate when the $\mathbf{a}_i(\mathbf{x}^*)$ are independent and that that will be assumed to be so.

(c) The condition (5.2) is equivalent to

$$\nabla_x L(\mathbf{x}^*, \boldsymbol{\lambda}^*) = 0, \tag{5.5}$$

L being the Lagrangian. Thus $L(\mathbf{x}, \boldsymbol{\lambda}^*)$ has a stationary point at $\mathbf{x}^*$. Condition (5.3) means

$$L(\mathbf{x}^*, \boldsymbol{\lambda}^*) = f(\mathbf{x}^*). \tag{5.6}$$

(d) Sufficient conditions for $\mathbf{x}^*$ to be a local minimum on the active set I are that, for some $\varepsilon > 0$, and for all $\|\mathbf{x} - \mathbf{x}^*\| \leqslant \varepsilon$,

$$\mathbf{p}^{\mathrm{T}}\mathbf{F}(\mathbf{x}, \boldsymbol{\lambda}^*)\mathbf{p} \geqslant \varepsilon \|\mathbf{p}\|^2 \tag{5.7}$$

for all $\mathbf{p}$ such that

$$\mathbf{A}\mathbf{p} = 0, \tag{5.8}$$

where $\mathbf{F}$ is the Hessian of the Lagrangian and $\mathbf{A}$ is the matrix whose rows are $\mathbf{a}_i^{\mathrm{T}}(\mathbf{x}^*)$, $i \in I$.

If f is convex, c_i concave, (5.2) and (5.3) are sufficient for a global minimum.

Note that the solution of a constrained problem involves both the identification of the active set and minimisation on that set.

5.2 Projection methods

In this section $c_i(\mathbf{x})$ is taken to be linear, all i: to keep the notation consistent,

$$c_i(\mathbf{x}) = \mathbf{a}_i^{\mathrm{T}}\mathbf{x} - b_i, \quad i = 1, 2, \ldots m. \tag{5.9}$$

With $\mathbf{x}_r$ the current point, I_r the current active set, define $\mathbf{A}_r$ as the $t \times n$ matrix whose rows are $\mathbf{a}_i^{\mathrm{T}}$, $i \in I_r$. The definition of I_r

$$I_r = \{i \mid c_i(\mathbf{x}_r) = 0\}$$

means here

$$\mathbf{A}_r\mathbf{x}_r = \mathbf{b}_r, \tag{5.10}$$

where $\mathbf{b}_r$ is the t-vector whose elements are b_i, $i \in I_r$. If now a move is made in the direction $\mathbf{p}$, so that

$$\mathbf{x}_{r+1} = \mathbf{x}_r + \alpha\mathbf{p},$$

then $\mathbf{x}_{r+1}$ still satisfies the active constraints if

$$\mathbf{A}_r\mathbf{p} = 0. \tag{5.11}$$

The idea of all projection methods is to use directions $\mathbf{p}$ satisfying (5.11). Some commonly used methods are described in the next section, where for convenience the suffix r is dropped.

5.2.1 Some projections

(a) Orthogonal projection Any vector $\mathbf{u}$ in R^n can be written as the sum of two vectors, $\mathbf{p}$ in the manifold defined by (5.11) and $\mathbf{q}$

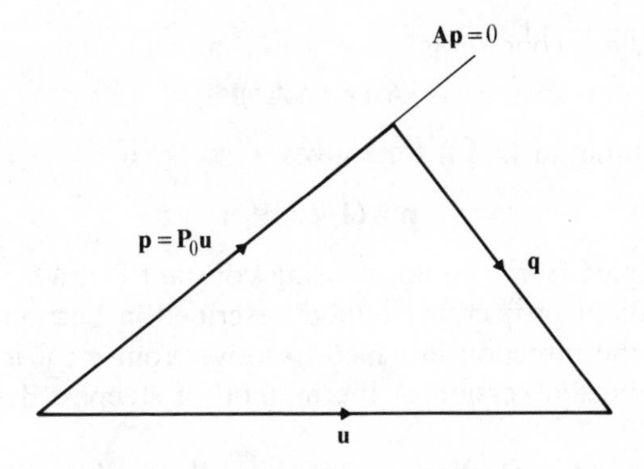

Fig. 5.1. Orthogonal projection.

orthogonal to this manifold, so that

$$\mathbf{q} = \mathbf{A}^{\mathrm{T}}\mathbf{z} \tag{5.12}$$

where $\mathbf{z} \in R^t$. This is obvious by considering the bases of $\mathbf{p}$ and $\mathbf{q}$, or simply by finding the appropriate representation; if

$$\mathbf{u} = \mathbf{p} + \mathbf{q} = \mathbf{p} + \mathbf{A}^{\mathrm{T}}\mathbf{z},$$

then from (5.11)

$$\mathbf{Au} = (\mathbf{AA}^{\mathrm{T}})\mathbf{z},$$

and since $\mathbf{AA}^{\mathrm{T}}$ is nonsingular,

$$\mathbf{z} = (\mathbf{AA}^{\mathrm{T}})^{-1}\mathbf{Au} \tag{5.13}$$

so that

$$\mathbf{q} = \mathbf{A}^{\mathrm{T}}(\mathbf{AA}^{\mathrm{T}})^{-1}\mathbf{Au}, \quad \mathbf{p} = \{\mathbf{I} - \mathbf{A}^{\mathrm{T}}(\mathbf{AA}^{\mathrm{T}})^{-1}\mathbf{A}\}\mathbf{u} = \mathbf{P}_0\mathbf{u}. \tag{5.14}$$

Clearly $\mathbf{P}_0\mathbf{q} = 0$, $\mathbf{P}_0\mathbf{p} = \mathbf{p}$ and $\mathbf{P}_0\mathbf{P}_0 = \mathbf{P}_0$ so that $\mathbf{P}_0$ is an orthogonal projection operator (see Fig. 5.1).

The projection $\mathbf{P}_0\mathbf{u}$ is also the vector closest to $\mathbf{u}$ in the sense that it maximises $\mathbf{u}^{\mathrm{T}}\mathbf{p}$ for a $\mathbf{p}$ of fixed length. This follows by considering the maximum of $\mathbf{u}^{\mathrm{T}}\mathbf{p}$ subject to (5.11) and to $\|\mathbf{p}\|^2 = p_0^2$, and forming the Lagrangian

$$L(\mathbf{p}, \mathbf{l}, w) = \mathbf{u}^{\mathrm{T}}\mathbf{p} - \mathbf{l}^{\mathrm{T}}\mathbf{Ap} - w(\mathbf{p}^{\mathrm{T}}\mathbf{p} - p_0^2). \tag{5.15}$$

Necessary conditions at a maximum are

$$\nabla_p L = \mathbf{u} - \mathbf{A}^{\mathrm{T}}\mathbf{l} - 2w\mathbf{p} = 0 \tag{5.16}$$

so that, using (5.11),

$$\mathbf{Au} = (\mathbf{AA}^{\mathrm{T}})\mathbf{l} \tag{5.17}$$

and substituting in (5.16) for $\mathbf{l}$ gives

$$\mathbf{p} = (1/2w)\mathbf{P}_0\mathbf{u}$$

as before apart from the normalising constant $(1/2w)$.

The gradient projection method described in Section 5.3.1 uses $\mathbf{P}_0(-\mathbf{g}_r)$ as the direction in which to move from $\mathbf{x}_r$; it is therefore just the projected version of the method of steepest descents.

(b) Generalised projection – reduced gradient A natural idea for minimising $f(\mathbf{x})$ subject to t equality constraints is to use variable reduction, that is to solve for t of the x_i in terms of the $(n-t)$ other variables, and to regard these t x_i as dependent; f is then a function of $(n-t)$ independent and unconstrained variables and the changes in these which minimise f produce consequent changes in the dependent x_i. Wolfe (1967) suggested calling the gradient of f with respect to the independent variables the "reduced gradient".

The idea of this can be generalised. Consider linear transformations between the n-vector $\mathbf{x}$ and $\mathbf{y}_1$, $\mathbf{y}_2$, respectively t- and $(n-t)$-vectors.

$$\begin{aligned} \mathbf{y}_1 &= \mathbf{Ax} - \mathbf{b}, \\ \mathbf{y}_2 &= \mathbf{Sx} - \mathbf{v}. \end{aligned} \tag{5.18}$$

Here $\mathbf{S}$ is an arbitrary $(n-t)\times n$ matrix with the proviso only that

$$\mathbf{T} = \left(\frac{\mathbf{A}}{\mathbf{S}}\right)$$

is to be nonsingular, so that the relations (5.18) can be written

$$\mathbf{y} = \mathbf{Tx} - \mathbf{K}$$

and solved to give

$$\mathbf{x} = \mathbf{T}^{-1}\mathbf{y} + \mathbf{T}^{-1}\mathbf{K}. \tag{5.19}$$

Feasible changes in $\mathbf{x}$ then correspond to $\mathbf{y}_1 = 0$ and so are in the direction $\mathbf{p} = \mathbf{Zv}$, where $\mathbf{Z}$ is the matrix containing the last $n-t$ columns of $\mathbf{T}^{-1}$ and $\mathbf{v}$ is a $(n-t)$-vector of changes in $\mathbf{y}_2$. Since

$$f(\mathbf{x}) = f(\mathbf{T}^{-1}\mathbf{y} + \mathbf{T}^{-1}\mathbf{K}), \tag{5.20}$$

the reduced gradient $\nabla_{y_2} f$ is $\mathbf{Z}^{\mathrm{T}}\mathbf{g}$, and so $\mathbf{y}_2$ should be changed so that

$$\mathbf{v} = -\mathbf{Z}^{\mathrm{T}}\mathbf{g},$$

and as above

$$\mathbf{p} = -\mathbf{Z}\mathbf{Z}^{\mathrm{T}}\mathbf{g} = \mathbf{P}_R(-\mathbf{g}). \tag{5.21}$$

Note that this does not project into itself, that is $\mathbf{P}_R\mathbf{P}_R \neq \mathbf{P}_R$.

Any vector $\mathbf{p}$ satisfying $\mathbf{Ap} = 0$ has the columns of $\mathbf{Z}$ as a basis, and so can be written

$$\mathbf{p} = \mathbf{Zv}.$$

This follows immediately since

$$\mathbf{T} = \left(\frac{\mathbf{A}}{\mathbf{S}}\right), \quad \mathbf{T}^{-1} = (\mathbf{B} \mid \mathbf{Z}), \quad \text{so that } \mathbf{AZ} = 0. \tag{5.22}$$

It is easy to confirm that the orthogonal projection $\mathbf{P}_0\mathbf{u}$ can be written

$$\mathbf{P}_0\mathbf{u} = \mathbf{Z}(\mathbf{Z}^{\mathrm{T}}\mathbf{Z})^{-1}\mathbf{Z}^{\mathrm{T}}\mathbf{u} \tag{5.23}$$

by noting that $\mathbf{u} = \mathbf{P}_0\mathbf{u} + \mathbf{A}^{\mathrm{T}}\mathbf{z} = \mathbf{Zv} + \mathbf{A}^{\mathrm{T}}\mathbf{z}$ and multiplying through by $\mathbf{Z}^{\mathrm{T}}$. The behaviour of this method is dependent on the choice of $\mathbf{S}$ and so of $\mathbf{Z}$. A simple example is given in Section 5.3.2.

(c) Projected Newton direction Just as (a) gives the projected version of steepest descents, the vector "nearest" to $(-\mathbf{g})$ in the manifold $\mathbf{Ap} = 0$, so there is a projected version of Newton's method found by minimising

$$f(\mathbf{x}) = \tfrac{1}{2}\mathbf{x}^{\mathrm{T}}\mathbf{G}\mathbf{x} - \mathbf{b}^{\mathrm{T}}\mathbf{x}$$

for $\mathbf{x} = \mathbf{x}_r + \mathbf{p}$, as in (2.39), but now subject to $\mathbf{Ap} = 0$. Using the Lagrangian

$$L(\mathbf{p}, \boldsymbol{\lambda}) = \tfrac{1}{2}(\mathbf{x}_r + \mathbf{p})^{\mathrm{T}}\mathbf{G}(\mathbf{x}_r + \mathbf{p}) - \mathbf{b}^{\mathrm{T}}(\mathbf{x}_r + \mathbf{p}) - \boldsymbol{\lambda}^{\mathrm{T}}\mathbf{Ap}, \tag{5.24}$$

the necessary condition for a minimum is

$$\mathbf{G}\mathbf{x}_r - \mathbf{b} + \mathbf{Gp} - \mathbf{A}^{\mathrm{T}}\boldsymbol{\lambda} = 0, \tag{5.25}$$

and since the gradient $\mathbf{g} = \mathbf{G}\mathbf{x}_r - \mathbf{b}$, and $\mathbf{G}$ is assumed nonsingular,

$$\mathbf{p} = \mathbf{G}^{-1}\mathbf{A}^{\mathrm{T}}\boldsymbol{\lambda} - \mathbf{G}^{-1}\mathbf{g}. \tag{5.26}$$

Now $\mathbf{A}\mathbf{G}^{-1}\mathbf{A}^{\mathrm{T}}$ is also nonsingular if $\mathbf{A}$ is of full rank and so substituting for $\boldsymbol{\lambda}$ gives the projected version of the unconstrained Newton step $-\mathbf{G}^{-1}\mathbf{g}$,

$$\mathbf{p}=\mathbf{P}_N(-\mathbf{G}^{-1}\mathbf{g})=\{\mathbf{I}-\mathbf{G}^{-1}\mathbf{A}^{\mathrm{T}}(\mathbf{A}\mathbf{G}^{-1}\mathbf{A}^{\mathrm{T}})^{-1}\mathbf{A}\}(-\mathbf{G}^{-1}\mathbf{g}). \quad (5.27)$$

This $\mathbf{P}_N$ reduces to $\mathbf{P}_0$ when $\mathbf{G}=\mathbf{I}$, and is a projection operator since $\mathbf{P}_N\mathbf{P}_N=\mathbf{P}_N$. It is used for quadratic programming as described in Section 5.3.3.

(d) Projected directions, quasi Newton methods Section 2.4 described methods of unconstrained minimisation using

$$\mathbf{p}=-\mathbf{H}\mathbf{g}$$

together with an updating formula for the matrix $\mathbf{H}$ designed to have quadratic termination and values for $\mathbf{H}$ which change from $\mathbf{I}$ initially to $\mathbf{G}^{-1}$ finally. It is a logical extension to look for projected versions of these. The first was described by Goldfarb (1969) based on Davidon's 1959 paper; in this a step is projected on to the active set manifold initially by $\mathbf{H}_1=\mathbf{P}_0$. Then using any of the matrix updating formulae, subsequent steps are still in the same manifold since $\mathbf{A}\mathbf{H}_r=0$ implies $\mathbf{A}\mathbf{H}_{r+1}=0$ (confirm this for the DFP formula). Also if the active set remains unchanged for $n-t$ steps, it can be shown that the minimisation has quadratic termination and that

$$\mathbf{H}_{n-t+1}=\mathbf{P}_N\mathbf{G}^{-1}.$$

The method involves extensive algebra and will not be described further here. It has proved efficient in practice, as have other quasi Newton methods using different ways of updating which estimate the Hessian directly rather than its inverse. See, e.g. Gill and Murray (1974).

(e) Estimates of Lagrange multipliers A minimum subject to an active constraint set occurs when $\|\mathbf{p}\|=0$ and is characterised by the Lagrange multipliers $\boldsymbol{\lambda}$ for which

$$\mathbf{g}=\mathbf{A}^{\mathrm{T}}\boldsymbol{\lambda}. \quad (5.28)$$

A necessary condition that this shall be a minimum over $\mathbf{x}\in X$ is that $\boldsymbol{\lambda}\geqslant 0$, so having found a point where $\|\mathbf{p}\|=0$, we need to test $\boldsymbol{\lambda}$. Estimates of $\boldsymbol{\lambda}$ are obtained during the calculations and you can confirm that

(a) $\boldsymbol{\lambda} = (\mathbf{A}\mathbf{A}^{\mathrm{T}})^{-1}\mathbf{A}\mathbf{g}$ (5.29)

(b) $\boldsymbol{\lambda} = \mathbf{B}^{\mathrm{T}}\mathbf{g}$, where $\mathbf{T}^{-1} = (\mathbf{B} \mid \mathbf{Z})$ (5.30)

(c) $\boldsymbol{\lambda} = (\mathbf{A}\mathbf{G}^{-1}\mathbf{A}^{\mathrm{T}})^{-1}\mathbf{A}\mathbf{G}^{-1}\mathbf{g}$. (5.31)

Equations (5.29), (5.30) may be regarded as first-order estimates of $\boldsymbol{\lambda}$, (5.31) as a second-order estimate since it includes information about the Hessian. If $\|\mathbf{p}\| = 0$ exactly, then of course these formulae are exact; numerical difficulties arise mainly from finding columns of $\mathbf{Z}$ nearly dependent.

5.2.2 Method of solution of inequality-constrained problems using projection

Use of any projection method involves the following sequence of steps, starting at the current feasible point $\mathbf{x}_r$ with active set I_r and matrix $\mathbf{A}_r$.

Step 1 Find $\mathbf{g}_r = \mathbf{g}(\mathbf{x}_r)$ and $\mathbf{p}_r$, its projection

$$\mathbf{p}_r = \mathbf{P}(-\mathbf{g}_r),$$

with $\mathbf{P}$ taken as the appropriate operator, e.g. (5.14), (5.21), (5.27). This defines the acceptable direction for improving f on the current active set. There are two possibilities, $\|\mathbf{p}_r\| = 0$ or $\|\mathbf{p}_r\| \neq 0$ (of course in practice this means $\|\mathbf{p}_r\| < \varepsilon$).

Case A, $\|\mathbf{p}_r\| = 0$. In this case f cannot be improved without changing I_r and the Lagrange multipliers must be tested.

Step 2 Evaluate $\boldsymbol{\lambda}_r$ using (5.29), (5.30) or (5.31). If all elements of $\boldsymbol{\lambda}_r$ are non-negative, then a minimum has been reached and the process stops. If some elements of $\boldsymbol{\lambda}_r$ are negative, drop from the active set the constraint corresponding to the smallest element and return to Step 1. This is the same process as is used in the simplex method, where constraints are adjusted one at a time.

Note that the new projected direction will automatically be feasible. When $\mathbf{p} = 0$ and $\lambda_t < 0$ it follows that

$$\mathbf{g} = \sum_{i=1}^{t-1} \lambda_i \mathbf{a}_i + \lambda_t \mathbf{a}_t,$$

and the new direction $\mathbf{p}'$ for which $\mathbf{a}_i^{\mathrm{T}}\mathbf{p}' = 0$, $i = 1, 2, \ldots, t-1$ must then satisfy

$$\mathbf{g}^{\mathrm{T}}\mathbf{p}' = \lambda_t \mathbf{a}_t^{\mathrm{T}}\mathbf{p}'. \tag{5.32}$$

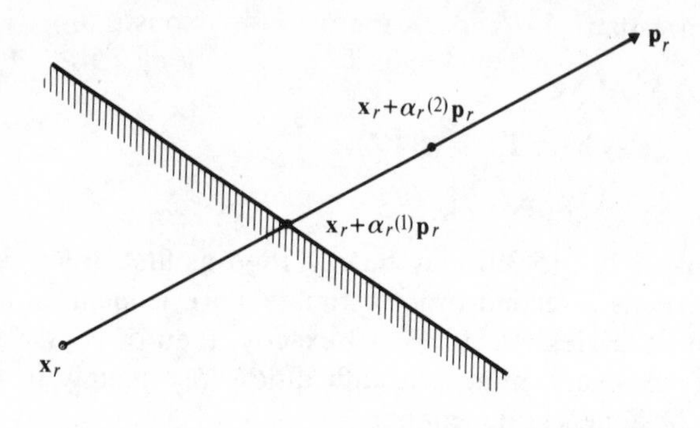

Fig. 5.2. Step ends either at minimum or on constraint.

Hence a $\mathbf{p}'$ which decreases f, so that $\mathbf{g}^{\mathrm{T}}\mathbf{p}'<0$, also increases $c_t(\mathbf{x})$ and so is a feasible direction.
Case B, $\|\mathbf{p}_r\|\neq 0$. A step in the direction $\mathbf{p}_r$ will improve f. Such a step can be ended in two ways, either by meeting a constraint which is currently inactive or by reaching the minimum of f along this direction – or, as a variant, simply getting to a smaller value of f than f_r. Suppose that the first case corresponds to

$$\mathbf{x}_{r+1}=\mathbf{x}_r+\alpha_r^{(1)}\mathbf{p}_r,$$

where $\alpha_r^{(1)}$ is the smallest $\alpha>0$ such that $c_i(\mathbf{x}_r+\alpha_r\mathbf{p}_r)=0$ for some $i\notin I_r$, and that

$$\mathbf{x}_{r+1}=\mathbf{x}_r+\alpha_r^{(2)}\mathbf{p}_r$$

minimises $f(\mathbf{x}_r+\alpha\mathbf{p}_r)$ with respect to α. Then choose α_r as the smaller of $\alpha_r^{(1)}$, $\alpha_r^{(2)}$ and the corresponding $\mathbf{x}_{r+1}$ as the new point.
Step 2 Return to Step 1 with $\mathbf{x}_{r+1}$ and an active set augmented by the new constraint if $\alpha=\alpha^{(1)}$ and unchanged if $\alpha=\alpha^{(2)}$.

The requirements for any algorithm of this type are a method of finding an initial feasible point and a suitable way of evaluating the projection operators and of changing them when constraints are added to or dropped from the active set.

Evaluation of projection operators To change an active set by one constraint only produces simple formulae involving bordered matrices. Thus, for instance, the orthogonal projection operator $\mathbf{P}_0^t$ with the first t constraints active becomes, on the addition of the $(t+1)$st constraint,

$$\mathbf{P}_0^{t+1}=\mathbf{P}_0^t-\mathbf{P}_0^t\mathbf{a}_{t+1}\mathbf{a}_{t+1}^{\mathrm{T}}\mathbf{P}_0^t/(\mathbf{a}_{t+1}^{\mathrm{T}}\mathbf{P}_0^t\mathbf{a}_{t+1}). \tag{5.33}$$

This can be used to build up $\mathbf{P}_0^t$ by adding constraints one at a time starting with $\mathbf{P}_0^0 = \mathbf{I}$. Similarly the deletion of the t-th constraint leads to

$$\{\mathbf{A}^{t-1}(\mathbf{A}^{t-1})^{\mathrm{T}}\}^{-1} = \mathbf{B}_1 - \mathbf{B}_2\mathbf{B}_3^{-1}\mathbf{B}_2^{\mathrm{T}}, \tag{5.34}$$

where $\mathbf{B}_1$, $\mathbf{B}_2$, $\mathbf{B}_3$ are the submatrices, respectively $(t-1)\times(t-1)$, $(t-1)\times 1$ and 1×1, such that

$$\{\mathbf{A}^t(\mathbf{A}^t)^{\mathrm{T}}\}^{-1} = \begin{bmatrix} \mathbf{B}_1 & \mathbf{B}_2 \\ \mathbf{B}_2^{\mathrm{T}} & \mathbf{B}_3 \end{bmatrix}. \tag{5.35}$$

(These formulae are derived in Problem 3.) Note that $\mathbf{A}\mathbf{A}^{\mathrm{T}}$ is often illconditioned and so its inverse must be calculated with precautions against round-off error. Methods using matrix decomposition have proved most suitable.

Initial feasible point An initial point for linear constraints can be found as in Phase 1 of the simplex method by adding surplus variables and reducing their sum to zero. However, when a projection method is to be used, the projection operators can be used also to start the calculation. Suppose that an arbitrary point $\mathbf{x}_0$ lies on or outside the first q constraints, violating at least one of them; so that

$$\mathbf{A}_q\mathbf{x}_0 \leqslant \mathbf{b}_q,$$

where $\mathbf{A}_q$ is the matrix whose rows are $\mathbf{a}_i^{\mathrm{T}}$, $1 \leqslant i \leqslant q$. Then a move $\mathbf{s}$ to a point $\mathbf{x}_1$ such that

$$\mathbf{A}_q\mathbf{x}_1 = \mathbf{b}_q$$

satisfies

$$\mathbf{A}_q\mathbf{s} = \mathbf{b}_q - \mathbf{A}_q\mathbf{x}_0 = \mathbf{r}_q \tag{5.36}$$

and so

$$\mathbf{x}_1 = \mathbf{x}_0 + \mathbf{A}_q^{\mathrm{T}}(\mathbf{A}_q\mathbf{A}_q^{\mathrm{T}})^{-1}\mathbf{r}_q. \tag{5.37}$$

This is a move orthogonal to the manifold M_q defined by $\mathbf{A}_q\mathbf{x} = 0$, and so the shortest step. The point $\mathbf{x}_1$ is then feasible for the first q constraints but may violate one or more of the others. Any new infeasibility can be removed by projecting its gradient along M_q; if this projection is zero, then a new manifold is chosen including this new constraint and dropping an existing one, and the point is projected orthogonally on to this. See Problem 2 for an example of this.

5.2.3 Projection methods – examples

Example 5.1 (Gradient projection.) Minimise

$$f(\mathbf{x})=(x_1-1)^2+(x_2-2)^2+(x_3-3)^2+(x_4-4)^2$$

subject to

$$c_1(\mathbf{x})\equiv -x_1-x_2-x_3-x_4+5\geqslant 0$$
$$c_2(\mathbf{x})\equiv -3x_1-3x_2-2x_3-x_4+10\geqslant 0$$
$$x_1\geqslant 0,\quad x_2\geqslant 0,\quad x_3\geqslant 0,\quad x_4\geqslant 0,$$

using the gradient projection method.

Choose as starting point $\mathbf{x}_1^T=(1/2, 1, 3/2, 2)$ for which $c_1(\mathbf{x})=0$.

$$\mathbf{A}_1=(-1\quad -1\quad -1\quad -1).$$

Step 1 $\mathbf{g}^T=(-1,-2,-3,-4)$

$$\mathbf{P}_0=\mathbf{I}-1/4\mathbf{A}_1^T\mathbf{A}_1 \quad \text{since} \quad (\mathbf{A}_1\mathbf{A}^T)^{-1}=1/4$$
$$\mathbf{p}_1=\mathbf{P}_0(-\mathbf{g}_1) \quad \text{so} \quad \mathbf{p}_1^T=(-3/2,-1/2,1/2,3/2), \text{ case B.}$$

Step 2 A move to $\mathbf{x}_1+\alpha\mathbf{p}_1$ stops on $x_1=0$ when $\alpha=1/3$, and minimises f with respect to α when $\alpha=1/2$. So $\alpha=1/3$ and

$$\mathbf{x}_2^T=(0, 5/6, 5/3, 5/2).$$

Constraints $c_1(\mathbf{x})\geqslant 0$ and $x_1\geqslant 0$ are active, so

$$\mathbf{A}_2=\begin{bmatrix}-1 & -1 & -1 & -1\\ 1 & 0 & 0 & 0\end{bmatrix}.$$

Step 1 $\mathbf{g}_2^T=(-2,-7/3,-8/3,-3)$

$$\mathbf{P}_0=\mathbf{I}-\mathbf{A}_2^T(\mathbf{A}_2\mathbf{A}_2^T)^{-1}\mathbf{A}_2$$
$$\mathbf{p}_2=\mathbf{P}_0(-\mathbf{g}_2), \quad \text{so} \quad \mathbf{p}_2^T=(0,-1/3,0,1/3), \text{ case B.}$$

Step 2 $\alpha_2^{(1)}=5/2$ for $x_2=0$; $\alpha_2^{(2)}=1/2$ for a minimum along $\mathbf{p}_2$, so $\alpha=1/2$.

$$\mathbf{x}_3^T=(0, 2/3, 5/3, 8/3) \text{ and } \mathbf{A}, \mathbf{P}_0 \text{ are unchanged.}$$

Step 1 $\mathbf{g}_3^T=(-2,-8/3,-8/3,-8/3)$

$$\mathbf{p}_3^T=0, \text{ case A.}$$

Step 2 The Lagrange multipliers are

$$\boldsymbol{\lambda}=(\mathbf{A}_2\mathbf{A}_2^T)^{-1}\mathbf{A}_2\mathbf{g}_3, \quad \text{so} \quad \boldsymbol{\lambda}^T=(8/3, 2/3).$$

These are non-negative so the point $\mathbf{x}_3$ satisfies necessary conditions for a minimum. Since f is convex, $\mathbf{c}$ linear, this is a global minimum.

Example 5.2 (Gradient projection.) Solve the same problem as Example 5.1 with the same method starting at $\mathbf{x}_1^{\mathrm{T}} = (0, 0, 0, 4)$.

This time

$$\mathbf{A}_1 = \begin{bmatrix} 1 & 0 & 0 & 0 \\ 0 & 1 & 0 & 0 \\ 0 & 0 & 1 & 0 \end{bmatrix}.$$

Step 1 $\mathbf{g}_1^{\mathrm{T}} = (-2, -4, -6, 0)$, $\mathbf{p}_1 = 0$, case A

Step 2 $\boldsymbol{\lambda}^{\mathrm{T}} = (-2, -4, -6)$ so the minimum is not reached. Drop the constraint corresponding to the most negative λ_i, so now use

$$\mathbf{A}_2 = \begin{bmatrix} 1 & 0 & 0 & 0 \\ 0 & 1 & 0 & 0 \end{bmatrix}$$

and $\mathbf{x}_2 = \mathbf{x}_1$, $\mathbf{g}_2 = \mathbf{g}_1$.

Step 1 $\mathbf{p}_2^{\mathrm{T}}(0, 0, 6, 0)$, case B.

Step 2 $\alpha^{(1)} = 1/6$ when $c_1(\mathbf{x}) = 0$, $\alpha^{(2)} = 1/2$. So $\mathbf{x}_3^{\mathrm{T}} = (0, 0, 1, 4)$ and

$$\mathbf{A}_3 = \begin{bmatrix} 1 & 0 & 0 & 0 \\ 0 & 1 & 0 & 0 \\ -1 & -1 & -1 & -1 \end{bmatrix}.$$

Step 1 $\mathbf{g}_3^{\mathrm{T}} = (-2, -4, -4, 0)$, $\mathbf{p}_3^{\mathrm{T}} = (0, 0, 2, -2)$ and the process continues as before, reaching the same minimum.

Example 5.3 (Reduced gradient). Solve the same problem as in Examples 5.1, 5.2 by the reduced gradient method.

Choose $\mathbf{x}_1^{\mathrm{T}} = (1/2, 1, 3/2, 2)$ for which $c_1(\mathbf{x}) = 0$ is the only active *constraint.*

Step 1 Write $y_1 = -x_1 - x_2 - x_3 - x_4 + 5$ and take for the other elements of $\mathbf{y}$ any linear functions of the x's. Here use variable reduction and take simply

$$y_2 = x_1, \quad y_3 = x_2, \quad y_4 = x_3,$$

and so

$$x_4 = -y_1 - y_2 - y_3 - y_4 + 5,$$

so that initially $\mathbf{y}^{\mathrm{T}} = (0, 1/2, 1, 3/2)$. The reduced gradient is

$$\partial f/\partial y_2, \partial f/\partial y_3, \partial f/\partial y_4, \quad \text{subject to } y_1 = 0,$$

and so is

$$\partial f/\partial x_1 - \partial f/\partial x_4, \partial f/\partial x_2 - \partial f/\partial x_4, \partial f/\partial x_3 - \partial f/\partial x_4.$$

Initially the reduced gradient is $\mathbf{g}_R^T = (3, 2, 1)$, so the new $\mathbf{y}^T$ is $(0, 1/2 - 3\alpha, 1 - 2\alpha, 3/2 - \alpha)$ and the corresponding point is

$$\mathbf{x}_2^T = (1/2 - 3\alpha, 1 - 2\alpha, 3/2 - \alpha, 2 + 6\alpha);$$

that is,

$$\mathbf{p}_1^T = (-3, -2, -1, 6).$$

This can be calculated directly from 5.2.1(b) as

$$\mathbf{p}_1 = \mathbf{P}_R(-\mathbf{g}_1) = \mathbf{Z}\mathbf{Z}^T(-\mathbf{g}_1),$$

where $\mathbf{Z}$ is the last 3 columns of $\mathbf{T}^{-1}$, $\mathbf{T}$ is the matrix

$$\mathbf{T} = \begin{bmatrix} -1 & -1 & -1 & -1 \\ \hline 1 & 0 & 0 & 0 \\ 0 & 1 & 0 & 0 \\ 0 & 0 & 1 & 0 \end{bmatrix}$$

and $\mathbf{g}_1^T = (-1, -2, -3, -4)$.

Step 2 Using $\mathbf{p}_1$ as above, $\alpha^{(1)} = 1/6$, $\alpha^{(2)} = 0.38$ so $\alpha = 1/6$, $\mathbf{x}_2^T = (0, 2/3, 4/3, 3)$ and $y_2 = 0$ is a new active constraint. Now $\mathbf{y}_2^T = (0, 0, 2/3, 4/3)$.

Step 1 $\mathbf{g}_2^T = (-2, -8/3, -10/3, -2)$, and $\mathbf{Z}$ is now the last two columns of $\mathbf{T}^{-1}$, giving

$$\mathbf{p}_2 = \mathbf{P}_R(-\mathbf{g}_2) = (0, 2/3, 4/3, -2)$$

Step 2 f is minimised at

$$\mathbf{x}_3^T = (0, 0.786, 1.571, 2.643)$$

and the active set remains unchanged until the point

$$\mathbf{x}^T = (0, 2/3, 5/3, 8/3),$$

$$\mathbf{g}^T = (-2, -8/3, -8/3, -8/3) \text{ and } \mathbf{P}_R(-\mathbf{g}) = 0.$$

Now

$$\boldsymbol{\lambda} = \mathbf{B}^T\mathbf{g},$$

where $\mathbf{B}$ is the first two columns of $\mathbf{T}^{-1}$, and this gives $\boldsymbol{\lambda}^T = (8/3, 2/3)$ as before.

5.3 Quadratic programming

The problem now is to minimise a quadratic function

$$f(\mathbf{x}) = \tfrac{1}{2}\mathbf{x}^{\mathrm{T}}\mathbf{G}\mathbf{x} - \mathbf{c}^{\mathrm{T}}\mathbf{x}$$

subject to $\mathbf{Ax} \geqslant \mathbf{b}$; we assume that f actually has a minimum on the feasible set.

The first method suggested for this, Wolfe's method, satisfied the Kuhn–Tucker conditions by a modified simplex procedure, but involved introducing extra variables and so increasing the dimensionality of the problem. This is described in Section 5.3.1, and a more practicable method for large-scale problems, using projection and Newton's method, is given in Section 5.3.2. Of course any of the projection methods of the last section could also be used, indeed Examples 5.1–3 are actually quadratic problems, for simplicity, but it is generally better to exploit the special structure of the quadratic problem.

5.3.1 Modified simplex method

Suppose that the problem is if necessary modified so that $\mathbf{x} \geqslant 0$. Then the Kuhn–Tucker conditions at a minimum are

$$\begin{aligned} \mathbf{G}\mathbf{x}^* - \mathbf{c} - \mathbf{A}^{\mathrm{T}}\boldsymbol{\lambda}^* + \boldsymbol{\mu}^* &= 0 \\ \boldsymbol{\lambda}^{*\mathrm{T}}(\mathbf{A}\mathbf{x}^* - \mathbf{b}) &= 0 \\ \boldsymbol{\mu}^{*\mathrm{T}}\mathbf{x}^* &= 0 \\ \boldsymbol{\mu}^* \geqslant 0, \quad \boldsymbol{\lambda}^* \geqslant 0, \quad \mathbf{A}\mathbf{x}^* \geqslant \mathbf{b}, \quad \mathbf{x}^* \geqslant 0 & \end{aligned} \tag{5.38}$$

or, introducing slack variables $\mathbf{y} = \mathbf{Ax} - \mathbf{b}$, we need to solve

$$\left.\begin{aligned} \mathbf{G}\mathbf{x}^* - \mathbf{A}^{\mathrm{T}}\boldsymbol{\lambda}^* + \boldsymbol{\mu}^* \qquad &= \mathbf{c} \\ \mathbf{A}\mathbf{x}^* \qquad\qquad -\mathbf{y}^* &= \mathbf{b} \end{aligned}\right\} \tag{5.39}$$

$$\left.\begin{aligned} \boldsymbol{\lambda}^{*\mathrm{T}}\mathbf{y}^* &= 0 \\ \boldsymbol{\mu}^{*\mathrm{T}}\mathbf{x}^* &= 0 \\ \boldsymbol{\lambda}^*, \boldsymbol{\mu}^*, \mathbf{x}^*, \mathbf{y}^* &\geqslant 0 \end{aligned}\right\} \tag{5.40}$$

The eqs. (5.39) are linear in the variables, and so a solution can be found by a phase 1 simplex procedure introducing non-negative surplus variables into each equation and reducing their sum to zero. The only extra restriction is that, because of (5.40), the choice of variable to join the basis is restricted; λ_i and y_i cannot both be in, and neither can μ_j, x_j.

Example 5.4 (Simplex method for quadratic programming.) Solve the quadratic problem of Example 5.1 using the restricted simplex method.

Equations (5.39) reduce to

$$\begin{aligned}
2x_1 \quad\quad + \lambda_1 + 3\lambda_2 - \mu_1 \quad\quad\quad\quad + w_1 &= 2\\
2x_2 \quad + \lambda_1 + 3\lambda_2 \quad - \mu_2 \quad\quad\quad + w_2 &= 4\\
2x_3 + \lambda_1 + 2\lambda_2 \quad\quad - \mu_3 \quad\quad + w_3 &= 6\\
2x_4 + \lambda_1 + \lambda_2 \quad\quad\quad - \mu_4 \quad + w_4 &= 8\\
x_1 + x_2 + x_3 + x_4 \quad\quad\quad\quad + y_1 &= 5\\
3x_1 + 3x_2 + 2x_3 + x_4 \quad\quad\quad\quad + y_2 &= 10
\end{aligned}$$

and minimising $w_1 + w_2 + w_3 + w_4$ gives a first tableau as shown in Table 5.1.

Table 5.1

w_1	2	0	0	0	1	3	−1	0	0	0	0	0	1	0	0	0	2
w_2	0	2	0	0	1	3	0	−1	0	0	0	0	0	1	0	0	4
w_3	0	0	2	0	1	2	0	0	−1	0	0	0	0	0	1	0	6
w_4	0	0	0	2	1	1	0	0	0	−1	0	0	0	0	0	1	8
y_1	1	1	1	1	0	0	0	0	0	0	1	0	0	0	0	0	5
y_2	3	3	2	1	0	0	0	0	0	0	0	1	0	0	0	0	10
	2	2	2	2	4	9	−1	−1	−1	−1	0	0	0	0	0	0	20

This tableau has 16 variables. Solving by the simplex method we note that λ_2 cannot enter the basis, nor λ_1, since y_1 and y_2 are both basic; hence the first pivot can be taken to replace w_1 by x_1; after several tableaux the final solution is found to be $\mathbf{w} = 0$ and then

$$\mathbf{x}^{*T} = (0, 2/3, 5/3, 8/3), \quad \boldsymbol{\lambda}^{*T} = (8/3, 0),$$
$$\boldsymbol{\mu}^{*T} = (2/3, 0, 0, 0), \quad \mathbf{y}^{*T} = (0, 6).$$

5.3.2 Projected Newton method

In Section 5.2.1(c) it was shown that the minimum of a quadratic function f subject to equality constraints $\mathbf{Ax} = \mathbf{b}$ is attained at the point when

$$\mathbf{x} = \mathbf{x}_r + \mathbf{p}$$

when $\mathbf{Ax}_r = \mathbf{b}$ and

$$\mathbf{p} = \mathbf{P}_N(-\mathbf{G}^{-1}\mathbf{g}_r) = \{\mathbf{I} - \mathbf{G}^{-1}\mathbf{A}^T(\mathbf{AG}^{-1}\mathbf{A}^T)^{-1}\mathbf{A}\}(-\mathbf{G}^{-1}\mathbf{g}_r).$$

The projected Newton method differs from the other projected methods only in that $\mathbf{x}_r+\mathbf{p}$ is the minimum of f in that direction and so $\alpha^{(2)}=1$ in Step 2. The inverse of $\mathbf{G}$ is required, and also of $\mathbf{A}\mathbf{G}^{-1}\mathbf{A}^{\mathrm{T}}$, and formulae are available for updating this when a constraint is added or removed, see Problem 8.

Example 5.5 (Newton's method, quadratic programming.) Minimise $f(\mathbf{x})=\frac{1}{2}\mathbf{x}^{\mathrm{T}}\mathbf{G}\mathbf{x}-\mathbf{c}^{\mathrm{T}}\mathbf{x}$, where

$$\mathbf{G}=\begin{bmatrix} 2 & -1 & -2 \\ -1 & 5 & -6 \\ -2 & -6 & 12 \end{bmatrix}, \quad \mathbf{c}=\begin{bmatrix} +6 \\ 0 \\ -10 \end{bmatrix},$$

and the constraints are $\mathbf{x}\geqslant 0$, $x_1+x_2+x_3\leqslant 5$. Take as initial point $\mathbf{x}_1^{\mathrm{T}}=(1,4,0)$ with two active constraints and

$$\mathbf{A}_1=\begin{bmatrix} 0 & 0 & 1 \\ -1 & -1 & -1 \end{bmatrix}.$$

Step 1 $\mathbf{g}_1^{\mathrm{T}}=(-8,19,-16)$, $\mathbf{p}_1^{\mathrm{T}}=(3,-3,0)$.

Step 2 $\alpha_1^{(1)}=4/3$, $\alpha_1^{(2)}=1$, so $\mathbf{x}_2=\mathbf{x}_1+\mathbf{p}_1$, $\mathbf{x}_2^{\mathrm{T}}=(4,1,0)$, and this point is a local minimum. The Lagrange multipliers from (5.40) are

$$\boldsymbol{\lambda}_2^{\mathrm{T}}=(-5,-1).$$

Drop the constraint corresponding to -5, that is $x_3=0$ and return to Step 1 with $\mathbf{x}_2$ and $\mathbf{A}_2=(-1\ -1\ -1)$.

Step 1 $\mathbf{g}_2^{\mathrm{T}}=(1,1,4)$, $\mathbf{p}_2^{\mathrm{T}}=(-50/161, 5/161, 45/161)$.

Step 2 $\alpha_2^{(1)}=644/50$, $\alpha_2^{(2)}=1$ so $\mathbf{x}_3^{\mathrm{T}}=(3.689, 1.031, 0.280)$. $\lambda=0.211$ so this is the minimum point.

5.4 Application of projection methods to nonlinear constraints

The idea of projection can be extended to this case by linearising the active constraints about the current point $\mathbf{x}_r$, that is by constructing the active set matrix $\mathbf{A}_r$ from the current gradients $\mathbf{a}_i(\mathbf{x}_r)$. One disadvantage is that new $\mathbf{a}$-vectors have to be calculated for each point and so the whole projection operator has to be worked out afresh at each step, even when the active set is unchanged. Also with a convex problem a move projected on to the manifold defined by the tangents to the constraints will be to

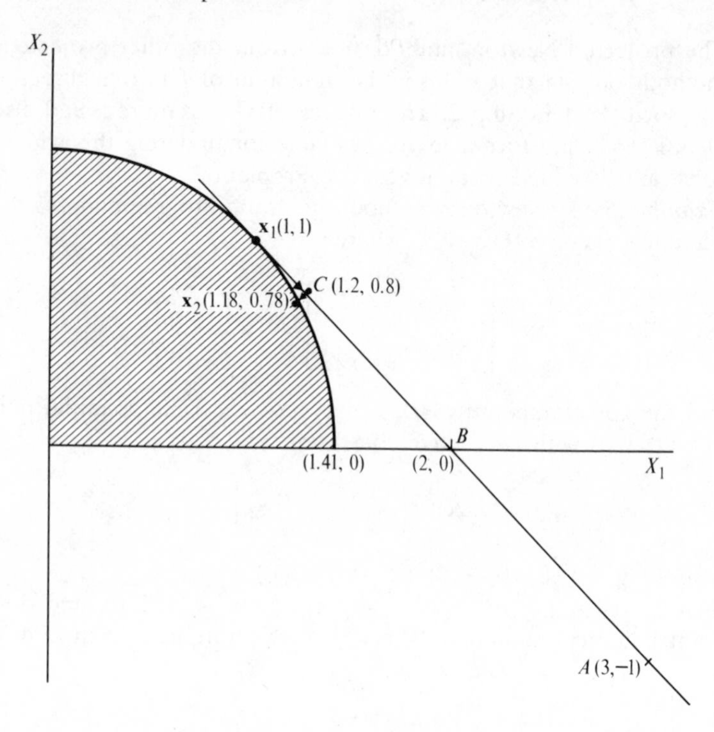

Fig. 5.3. Projection, nonlinear constraint.

an infeasible point; and a projection back on to the active constraints may produce a point at which the value of f is not decreased, or which is infeasible with respect to one or more of the other constraints.

Example 5.6 Minimise

$$f(\mathbf{x}) = 3(x_1 - 3)^2 + 2x_2^2 + x_1 x_2$$

subject to $-x_1^2 - x_2^2 \geqslant -2$, $x_1 \geqslant 0$, $x_2 \geqslant 0$, starting at $\mathbf{x}_1^T = (1, 1)$. Proceeding along $(+1, -1)$, the projection of the negative gradient $\mathbf{g}_1(+11, -5)$ along the tangent, gives a minimum at $(3, -1)$ which is infeasible. Stopping at the new constraint, $x_2 \geqslant 0$, produces the point $(2, 0)$ which is still infeasible with respect to the original constraint. Stopping at an earlier point, say $(1.2, 0.8)$, and then taking an orthogonal step to satisfy the first constraint, produces a feasible point $(1.18, 0.78)$ at which the value of f is smaller than the starting value.

An alternative formulation (Zoutendijk 1960) is to find a direction $\mathbf{p}$ which both maintains feasibility and decreases f, by solving at each step the linear programming problem.

maximise s subject to

$\mathbf{a}_i^{\mathrm{T}}\mathbf{p} \geqslant s$ for *nonlinear* active constraints i

$\mathbf{a}_i^{\mathrm{T}}\mathbf{p} = 0$ for linear active constraints or equality constraints

$\mathbf{g}^{\mathrm{T}}\mathbf{p} \leqslant -s$

$\mathbf{p}^{\mathrm{T}}\mathbf{p} = 1.$

Applying this to the above example with the same starting point gives

$$-p_1 - p_2 \geqslant s$$
$$-11p_1 + 5p_2 \leqslant -s$$
$$p_1^2 + p_2^2 = 1$$

and maximising s produces the values $p_1 = -1/\surd 10$, $p_2 = -3/\surd 10$,

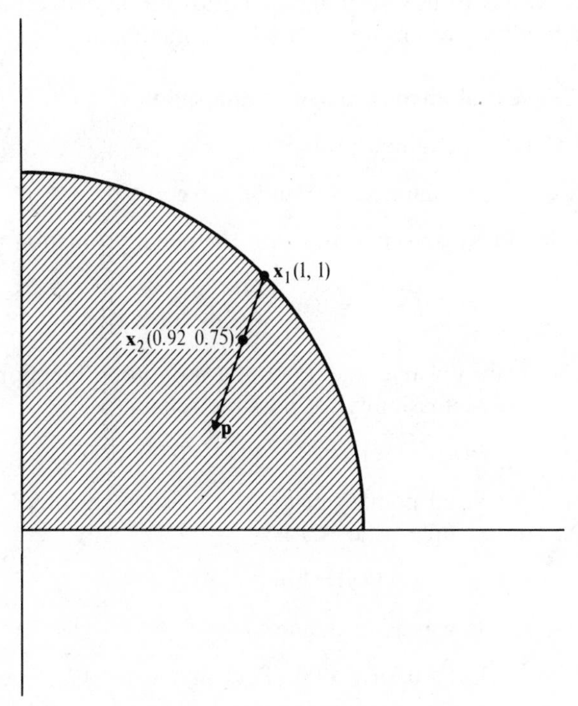

Fig. 5.4. Feasible directions.

$s = 4/\sqrt{10}$. This vector points into the interior of the feasible region and a step along it finds a minimum at the point $\mathbf{x}_2 = (0.917, 0.75)$. The next step depends only on $\mathbf{g}_2$ and will point back to the boundary of the feasible region.

5.5 Penalty function and multiplier methods

The idea of these methods is to solve a constrained minimisation problem by solving a sequence of unconstrained minimisation problems. A modified objective function has to be defined which in some way incorporates the constraints.

The simplest way of doing this is to ensure that this new objective function becomes large when the constraints are violated, by adding terms which have this behaviour, multiplied by weighting factors which are successively reduced as the calculation proceeds. The sequence of unconstrained minima thus produced tend under certain conditions to the constrained minimum point $\mathbf{x}^*$. Another way is to use the Lagrangian as the objective function, suitably modified to ensure that it has a minimum.

5.5.1 Sequential unconstrained minimisation

For the usual constrained problem

$$\text{minimise } f(\mathbf{x}) \text{ subject to } \mathbf{c}(\mathbf{x}) \geqslant 0,$$

define a modified objective function, or penalty function

$$\bar{f}(\mathbf{x}, \mathbf{r}) = f(\mathbf{x}) + \sum_{i=1}^{m} r_i P\{c_i(\mathbf{x})\}, \tag{5.41}$$

where the r_i are positive and $P(y)$ is a monotone function of y which penalises infeasibility. A possible scheme is

$$P(y) > 0 \text{ for } y < 0, \quad P(y) = 0 \text{ for } y \geqslant 0, \tag{5.42}$$

and then $\bar{f} > f$ at all points outside the feasible region X, $\bar{f} = f$ inside X. An example of (5.42) is

$$P(y) = \{\min(0, y)\}^2. \tag{5.43}$$

Another possible way is to define

$$P(y) > 0 \text{ for } y \geqslant 0, \quad P(y) \to \infty \text{ as } y \to 0^+. \tag{5.44}$$

In this case $\bar{f}$ increases as $\mathbf{x}$ approaches the boundary of X from the inside. Examples of this are

$$P(y) = -\ln y, \tag{5.45}$$

$$P(y) = 1/y. \tag{5.46}$$

A method using functions with the behaviour (5.44) is sometimes called a barrier-function method since the boundary of X provides an infinite barrier, and the unconstrained minima are all in the interior of X.

If $\bar{f}$, defined by (5.41), has a minimum at $\mathbf{x}(\mathbf{r})$, and if a sequence of $\mathbf{r}$, $\mathbf{r}_k$, are defined such that

$$\sum_{i=1}^{m} r_{k_i} P[c_i\{\mathbf{x}(\mathbf{r}_k)\}] \to 0 \quad \text{as } k \to \infty, \tag{5.47}$$

then clearly

$$\bar{f}\{\mathbf{x}(\mathbf{r}_k), \mathbf{r}_k\} \to f(\mathbf{x}^*) \quad \text{as } k \to \infty. \tag{5.48}$$

The great advantage of this is that it allows the use of the unconstrained minimisation algorithms and that nonlinear constraints need no special treatment. There are numerical difficulties, however, which will be mentioned later. Some examples will be given first, using the forms (5.43), (5.45), (5.46) and, for simplicity, taking $r_{k_i} = r_k$, all i, with $r_k \to 0$ as $k \to \infty$.

Example 5.7 Investigate minimising $f(x) = (x+1)^2$ subject to $x \geqslant 0$ using the three forms of penalty function.

Using (a) $\bar{f}(x, r) = (x+1)^2 + (1/r)\,\{\min(0, x)\}^2$ (b) $\bar{f}(x, r) = (x+1)^2 - r \ln x$ (c) $\bar{f}(x, r) = (x+1)^2 + r/x$ and $r = 1, 0.5, 0.1$, the behaviour of the three functions is sketched in Fig. 5.5. The minima of $\bar{f}$ are easy to obtain formally and are given in Table 5.2. The actual minimum is of course at $x^* = 0$.

Example 5.8 Minimise $f(\mathbf{x}) = 1 - x_1 + x_2$ subject to $c_1(\mathbf{x}) \equiv -x_1^2 + x_2 \geqslant 0$, $c_2(\mathbf{x}) \equiv x_1 \geqslant 0$.

Using form (a) the function $\bar{f}$ is

$$\bar{f}(\mathbf{x}, r) = 1 - x_1 + x_2 + (1/r)[\min\{0, c_1(\mathbf{x})\}]^2 + (1/r)[\min\{0, c_2(\mathbf{x})\}]^2.$$

In Fig. 5.6 are shown the zones A, B, C and the feasible region X in each of which the objective $\bar{f}$ has a different expression. Note that $\bar{f}$ is continuous but not differentiable on the boundaries between zones.

(b) $\bar{f}(\mathbf{x}, r) = 1 - x_1 + x_2 - r\{\ln(-x_1^2 + x_2) + \ln(x_1)\}$. Note that this function is only defined for $x \in X$, the solution set.

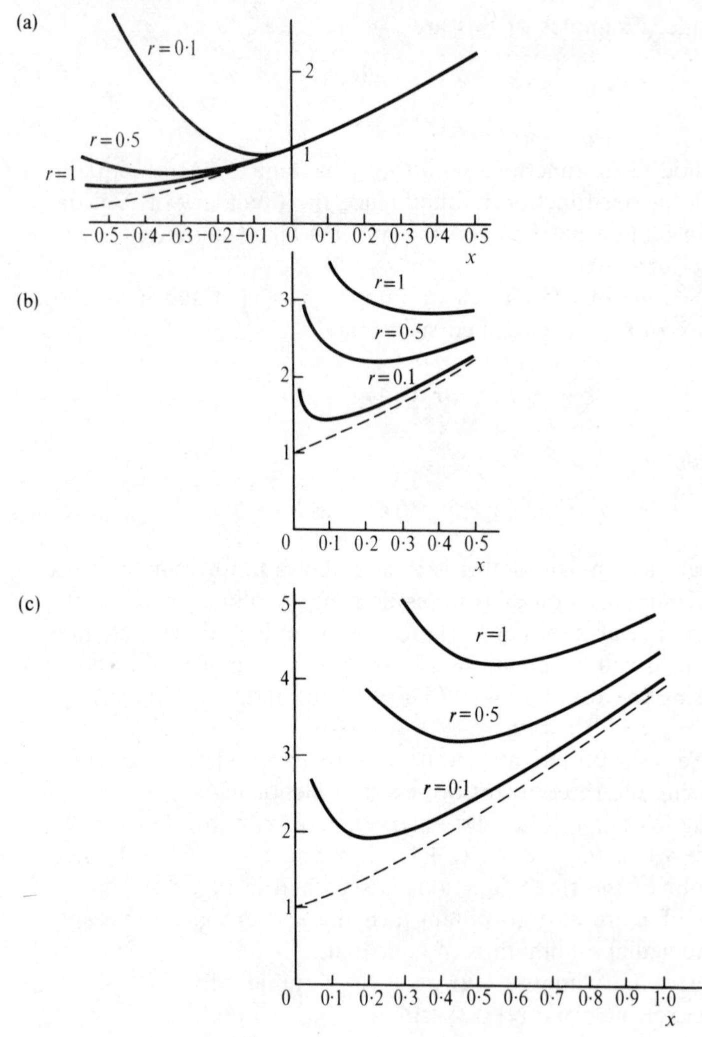

Fig. 5.5. Penalty functions for $f(x)=(x+1)^2$, $x \geqslant 0$.

Table 5.2

r	(a)	(b)	(c)
1	−0.5	0.366	0.565
0.5	−0.333	0.207	0.420
0.1	−0.091	0.048	0.206

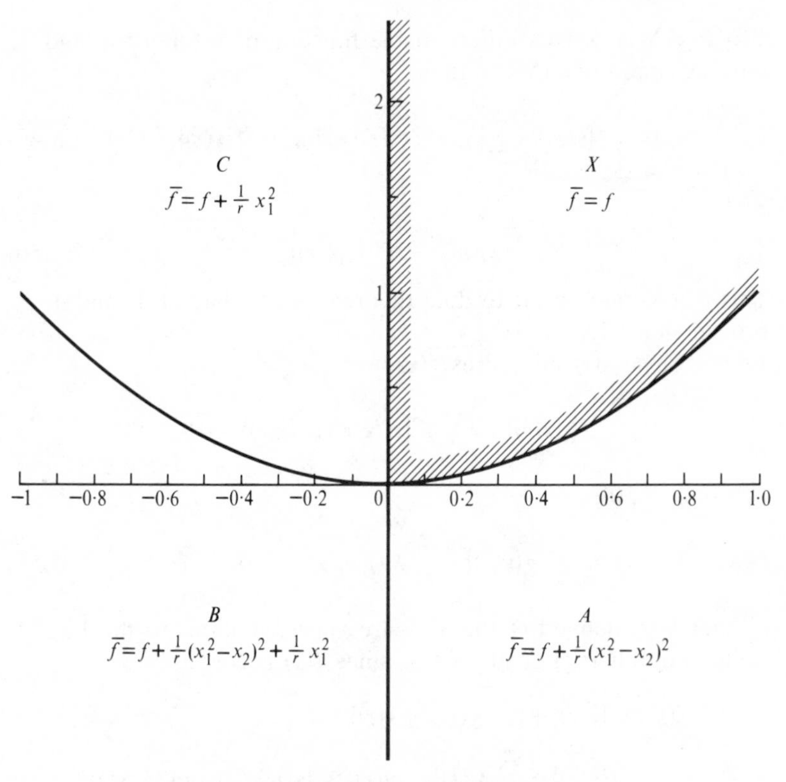

Fig. 5.6. Exterior point penalty function.

(c) $\bar{f}(\mathbf{x}, r) = 1 - x_1 + x_2 + r/(-x_1^2 + x_2) + r/x_1$.

The minima for $r = 1, 0.5, 0.1$ are now found to be as given in Table 5.3. All three methods tend to the minimum $\mathbf{x}^{*T} = (0.5, 0.25)$ and identify the active constraint c_1.

Bounds for interior-point algorithms Under certain special conditions interior-point methods like (b) and (c) can give bounds for $f(\mathbf{x}^*)$. If $f(\mathbf{x})$ is convex and the $c_i(\mathbf{x})$ concave, then X is convex; if

Table 5.3

r	(a)	(b)	(c)
1	(0.5, −0.25)	(1, 2)	(1, 2)
0.5	(0.5, 0)	(0.809, 1.154)	(0.848, 1.426)
0.1	(0.5, 0.2)	(0.585, 0.442)	(0.627, 0.709)

also $P(y)$ is a convex differentiable function of y for $y>0$, and satisfies conditions (5.44), then

$$f\{\mathbf{x}(r)\}-\sum_{i=1}^{m}\lambda_i(r)c_i\{\mathbf{x}(r)\}<f(\mathbf{x}^*)<f\{\mathbf{x}(r)\}, \tag{5.49}$$

where

$$\lambda_i(r)=-rP'[c_i\{\mathbf{x}(r)\}]. \tag{5.50}$$

This is a similar result to dual theorem A of Chapter 1, and is proved similarly.

Proof Since $\mathbf{x}(r)$ minimises $\bar{f}(\mathbf{x}, r)$ then

$$\mathbf{g}\{\mathbf{x}(r)\}+r\sum_{i=1}^{m}P'[c_i\{\mathbf{x}(r)\}]\mathbf{a}_i\{\mathbf{x}(r)\}=0$$

or using (5.50),

$$\mathbf{g}\{\mathbf{x}(r)\}-\sum_{i=1}^{m}\lambda_i(r)\mathbf{a}_i\{\mathbf{x}(r)\}=0 \tag{5.51}$$

so that $\lambda_i(r)$ defined in this way are approximations to the Lagrange multipliers. For any $\mathbf{x}\in X$, since $f(\mathbf{x})$ is convex

$$\begin{aligned} f(\mathbf{x}) &\geqslant f\{\mathbf{x}(r)\}+\{\mathbf{x}-\mathbf{x}(r)\}^{\mathrm{T}}\mathbf{g}\{\mathbf{x}(r)\} \\ &= f\{\mathbf{x}(r)\}+\sum_{i=1}^{m}\lambda_i(r)\{\mathbf{x}-\mathbf{x}(r)\}^{\mathrm{T}}\mathbf{a}_i\{\mathbf{x}(r)\} \quad \text{(from (5.51))} \\ &\geqslant f\{\mathbf{x}(r)\}+\sum_{i=1}^{m}\lambda_i(r)[c_i(\mathbf{x})-c_i\{\mathbf{x}(r)\}], \end{aligned}$$

since $c_i(\mathbf{x})$ is concave. Also $\lambda_i(r)\geqslant 0$ since $P'\leqslant 0$, $r>0$. Hence $\lambda_i(r)c_i(\mathbf{x})\geqslant 0$ for $\mathbf{x}\in X$, and so

$$f(\mathbf{x})\geqslant f\{\mathbf{x}(r)\}-\sum_{i=1}^{m}\lambda_i(r)c_i\{\mathbf{x}(r)\}=b(r)$$

for all $\mathbf{x}\in X$ and in particular for $\mathbf{x}=\mathbf{x}^*$, which proves one part of (5.49). The other part follows since $\mathbf{x}^*$ is the minimum of $f(\mathbf{x})$ for $\mathbf{x}\in X$ and so

$$f(\mathbf{x}^*)\leqslant f\{\mathbf{x}(r)\}.$$

These conditions hold for Example 5.7(b) and (c); the results for (b) give the bounds in Table 5.4, giving a range of [0.657, 0.857] which brackets the true value, 0.75.

Table 5.4

r	λ_1	λ_2	$f\{\mathbf{x}(r)\}$	$b(r)$
1	1	1	2	0
0.5	1	0.618	1.325	0.345
0.1	1	0.171	0.857	0.657

Numerical disadvantages The method requires repeated minimisation of $\bar{f}(\mathbf{x}, r_k)$ for a sequence of r_k; the result of each calculation, $\mathbf{x}(r_k)$, is used to provide a starting point for the next. In practice it is found that the behaviour of the method depends very critically on the way r_k tends to zero. If successive values of r_k differ too much, then the minimisations each take longer since the initial guess is not as good. However, small changes in r_k mean more minimisations must be done. The choice of the first r is also important. Fiacco and McCormick (1968) show that with $P(y) = -\ln y$, $\|\mathbf{x}(r) - \mathbf{x}^*\|$ is $O(r)$ while for $P(y) = 1/y$, $\|\mathbf{x}(r) - \mathbf{x}^*\|$ is $O(r^{1/2})$.

The performance of any of the minimisation algorithms becomes difficult as $\mathbf{x}(r)$ approaches the boundary of X, since with the interior point penalty functions $\bar{f}(\mathbf{x}, r)$ becomes infinite on the boundary and with the exterior point forms f has a discontinuous derivative there. This means that search methods based on fitted functions are difficult to apply. When the slope is very steep it may be impossible to reduce the magnitude of the gradient sufficiently to satisfy a test for convergence, since very large values of $\|\mathbf{g}\|$ can be attained close to the minimum, that is the Hessian matrix of $\bar{f}$ is increasingly ill-conditioned as r_k becomes small. All these difficulties arise from the fact that the function $\bar{f}$ becomes a more unsuitable shape as the method proceeds.

One way to deal with these difficulties is to extrapolate from the sequence $\mathbf{x}(r)$, that is to fit a smooth function to values of $\mathbf{x}(r)$ and take the value of this as $r \to 0$ as an estimate of $\mathbf{x}^*$. An example of this is given in Problem 15. Another method is to use the penalty function calculations only far enough to indicate the active constraints and then to minimise using these as equalities, that is to remove them from the penalty function. Finally the method given in Section 5.5.2 uses an augmented objective function whose added terms do not depend on a parameter which tends to zero in an arbitrary way but arise naturally from the Lagrange multipliers.

5.5.2 Lagrangian methods

The Lagrangian $L(\mathbf{x}, \boldsymbol{\lambda}^*)$ is known to have an unconstrained stationary point at $\mathbf{x}^*$, and $L(\mathbf{x}^*, \boldsymbol{\lambda}^*) = f(\mathbf{x}^*)$; so a natural way to define an augmented objective function might be thought to be to use the Lagrangian $L(\mathbf{x}, \boldsymbol{\lambda})$ with $\boldsymbol{\lambda}$ as some approximation to $\boldsymbol{\lambda}^*$. This can be expected to have a stationary point at $\mathbf{x}(\boldsymbol{\lambda})$, near $\mathbf{x}^*$. Unfortunately $L(\mathbf{x}, \boldsymbol{\lambda})$ may not have an unconstrained *minimum* at $\mathbf{x}(\boldsymbol{\lambda})$ – indeed even $L(\mathbf{x}, \boldsymbol{\lambda}^*)$ may not have an unconstrained minimum at $\mathbf{x}^*$. Consider for example the problem

$$\text{minimise } f(\mathbf{x}) = x_1^4 - x_1 x_2 + x_2 \quad \text{subject to } x_2 \geqslant 0.$$

The solution, by inspection, is $\mathbf{x}^{*\mathrm{T}} = (0, 0)$, $f(\mathbf{x}^*) = 0$. The Lagrangian is

$$L(\mathbf{x}, \boldsymbol{\lambda}) = x_1^4 - x_1 x_2 + x_2 - \lambda x_2,$$

and so $\nabla_{\mathbf{x}} L = 0$ gives $-x_1^* + 1 - \lambda^* = 0$, i.e. $\lambda^* = 1$. Then

$$L(\mathbf{x}, \lambda^*) = x_1^4 - x_1 x_2$$

and the Hessian of this is

$$\mathbf{F}(\mathbf{x}^*, \lambda^*) = \begin{bmatrix} 0 & -1 \\ -1 & 0 \end{bmatrix},$$

so that the second-order terms are certainly not positive for some feasible displacements from $(0, 0)$. The contours of $f(\mathbf{x})$ and of $L(\mathbf{x}, \lambda^*)$ are sketched in Fig. 5.7(a) and (b).

An ingenious way round this is to use an objective function $\hat{f}(\mathbf{x}, \boldsymbol{\lambda})$ produced by adding to the Lagrangian terms which ensure that $\hat{f}$ has an unconstrained minimum at $\mathbf{x}^*$ while maintaining the nice property that

$$\hat{f}(\mathbf{x}^*, \boldsymbol{\lambda}^*) = f(\mathbf{x}^*).$$

Alternate steps, first minimising $\hat{f}$ with respect to $\mathbf{x}$ for fixed $\boldsymbol{\lambda}$, and then updating $\boldsymbol{\lambda}$, finally converge to $\mathbf{x}^*$, $\boldsymbol{\lambda}^*$, and the minimising problems are well behaved. The proofs are sketched very briefly in the discussion below – a fuller treatment is given, for example, by Fletcher (1975).

Augmented Lagrangian methods for equality constraints It is convenient to describe the method first for equality constraints, so consider

$$\text{minimise } f(\mathbf{x}) \text{ subject to } \mathbf{c}(\mathbf{x}) = 0.$$

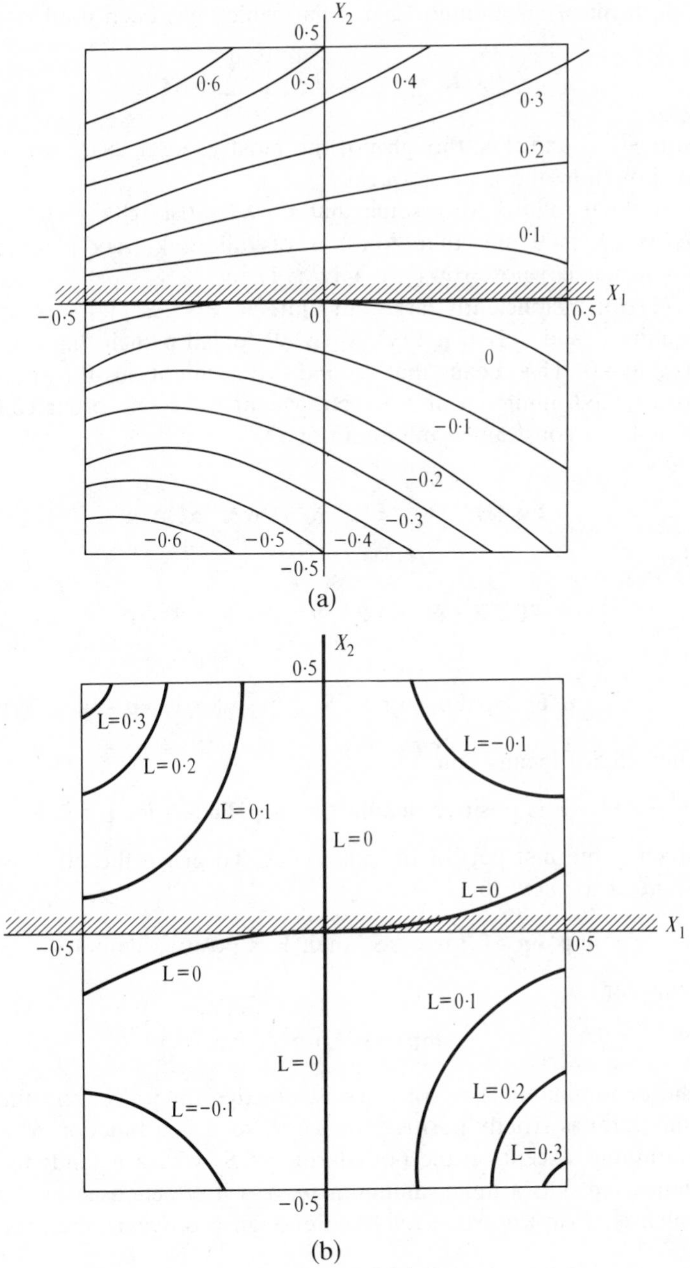

Fig. 5.7. (a) Contours of f (b) Contours of L.

One form of augmented Lagrangian which has been used is

$$\hat{f}(\mathbf{x}, \boldsymbol{\lambda}, w) = L(\mathbf{x}, \boldsymbol{\lambda}) + \tfrac{1}{2} w \sum_{i=1}^{m} \{c_i(\mathbf{x})\}^2$$

with $w > 0$ and this, though not the most general form, will serve to show how the process works.

In what follows we assume that $\mathbf{x}^*$, $\boldsymbol{\lambda}^*$ satisfy $c_i(\mathbf{x}^*) = 0$, $\nabla_{\mathbf{x}} L(\mathbf{x}^*, \boldsymbol{\lambda}^*) = 0$ and that $\mathbf{A}(\mathbf{x}^*)$ is of full rank, $\mathbf{A}(\mathbf{x}^*)$ being the $m \times n$ matrix whose rows are $\mathbf{a}_i^{\mathrm{T}}(\mathbf{x}^*)$. Then

(1) for a sufficiently large but finite w, $\hat{\mathbf{F}}(\mathbf{x}^*, \boldsymbol{\lambda}^*, w)$ is positive definite if and only if $\mathbf{p}^{\mathrm{T}}\mathbf{F}(\mathbf{x}^*, \boldsymbol{\lambda}^*)\,\mathbf{p} > 0$ for all $\mathbf{p}$ such that $\mathbf{A}(\mathbf{x}^*)\mathbf{p} = 0$. This means that second-order conditions for an unconstrained minimum of $\hat{f}$ are equivalent to second-order conditions for a constrained minimum of f.

Proof

$$\hat{\mathbf{F}}(\mathbf{x}^*, \boldsymbol{\lambda}^*, w) = \mathbf{F}(\mathbf{x}^*, \boldsymbol{\lambda}^*) + w\mathbf{A}^{\mathrm{T}}(\mathbf{x}^*)\mathbf{A}(\mathbf{x}^*).$$

Hence

$$\mathbf{p}^{\mathrm{T}}\hat{\mathbf{F}}\mathbf{p} = \mathbf{p}^{\mathrm{T}}\mathbf{F}\mathbf{p} \quad \text{for } \mathbf{p} \in S_f = \{\mathbf{p} \mid \|\mathbf{p}\| = 1, \mathbf{A}\mathbf{p} = 0\} \tag{5.52}$$

and

$$\mathbf{p}^{\mathrm{T}}\hat{\mathbf{F}}\mathbf{p} > \mathbf{p}^{\mathrm{T}}\mathbf{F}\mathbf{p} \quad \text{for } \mathbf{p} \in S_n = \{\mathbf{p} \mid \|\mathbf{p}\| = 1, \mathbf{A}\mathbf{p} \neq 0\}. \tag{5.53}$$

Then (5.52) means that

if $\hat{\mathbf{F}}$ is positive definite, then $\mathbf{p}^{\mathrm{T}}\mathbf{F}\mathbf{p} > 0$ for $\mathbf{p} \in S_f$,

which is the first part of the statement. To prove the other part, we need to show that

if $\mathbf{p}^{\mathrm{T}}\mathbf{F}\mathbf{p} > 0$ for $\mathbf{p} \in S_f$, then $\hat{\mathbf{F}}$ is positive definite.

Consider

$$\theta(\mathbf{p}) = \mathbf{p}^{\mathrm{T}}\mathbf{F}\mathbf{p}/\mathbf{p}^{\mathrm{T}}\mathbf{A}^{\mathrm{T}}\mathbf{A}\mathbf{p}. \tag{5.54}$$

The denominator is zero for $\mathbf{p} \in S_f$, positive for $\mathbf{p} \in S_n$, and the numerator is strictly positive for $\mathbf{p} \in S_f$, so θ is a function which is continuous except on the boundaries $\mathbf{p} \in S_f$, where it tends to $+\infty$. Hence $\theta(\mathbf{p})$ has a finite minimum m, and if we choose $-w < m$, which is always possible for large enough positive w, then for all $\mathbf{p}$

$$\mathbf{p}^{\mathrm{T}}\mathbf{F}\mathbf{p} > -w\mathbf{p}^{\mathrm{T}}\mathbf{A}^{\mathrm{T}}\mathbf{A}\mathbf{p}, \tag{5.55}$$

and this is

$$\mathbf{p}^{\mathrm{T}}\hat{\mathbf{F}}\mathbf{p}>0 \quad \text{for all } \mathbf{p} \tag{5.56}$$

as required.

(2) $\hat{f}(\mathbf{x}^*, \boldsymbol{\lambda}, w) = f(\mathbf{x}^*)$ since $c_i(\mathbf{x}^*) = 0$. (5.57)

(3) $\hat{f}(\mathbf{x}, \boldsymbol{\lambda}^*, w)$ has an unconstrained minimum at $\boldsymbol{\lambda}^*$: this follows from (1) and from the fact that

$$\nabla\hat{f}(\mathbf{x}^*, \boldsymbol{\lambda}^*, w) = \nabla L(\mathbf{x}^*, \boldsymbol{\lambda}^*) + w\mathbf{A}^{\mathrm{T}}(\mathbf{x}^*)\mathbf{c}(\mathbf{x}^*) = 0. \tag{5.58}$$

(4) $\hat{f}(\mathbf{x}, \boldsymbol{\lambda}, w)$ has an unconstrained minimum at $\mathbf{x}(\boldsymbol{\lambda}, w)$, near $\mathbf{x}^*$, for w sufficiently large – this follows by continuity from (1) and (3). Note that

$$\mathbf{x}(\boldsymbol{\lambda}^*, w) = \mathbf{x}^*, \tag{5.59}$$

and that

$$\nabla_{\mathbf{x}} f\{\mathbf{x}(\boldsymbol{\lambda}, w), \boldsymbol{\lambda}, w\} = \nabla_{\mathbf{x}} L\{\mathbf{x}(\boldsymbol{\lambda}, w), \boldsymbol{\lambda}\} + w\mathbf{A}^{\mathrm{T}}\{\mathbf{x}(\boldsymbol{\lambda}, w)\}\mathbf{c}\{\mathbf{x}(\boldsymbol{\lambda}, w)\} = 0. \tag{5.60}$$

(5) Defining a dual function, as in Chapter 1,

$$\phi(\boldsymbol{\lambda}, w) = \hat{f}\{\mathbf{x}(\boldsymbol{\lambda}, w), \boldsymbol{\lambda}, w\}, \tag{5.61}$$

we can show

(i) $\nabla_{\boldsymbol{\lambda}}\phi = -\mathbf{c}\{\mathbf{x}(\boldsymbol{\lambda}, w)\}$. (5.62)

Proof Differentiating (5.61),

$$\partial\phi/\partial\lambda_i = \nabla_{\mathbf{x}}\hat{f} - c_i\{\mathbf{x}(\boldsymbol{\lambda}, w)\} = -c_i\{\mathbf{x}(\boldsymbol{\lambda}, w)\} \quad \text{(from (5.60))}.$$

(ii) $\nabla^2_{\boldsymbol{\lambda}}\phi = -\mathbf{A}\{\mathbf{x}(\boldsymbol{\lambda}, w)\}[\hat{\mathbf{F}}\{\mathbf{x}(\boldsymbol{\lambda}, w), \boldsymbol{\lambda}, w\}]^{-1}\mathbf{A}^{\mathrm{T}}\{\mathbf{x}(\boldsymbol{\lambda}, w)\}$. (5.63)

Proof Differentiating (5.62) gives

$$\nabla^2_{\boldsymbol{\lambda}}\phi = -\mathbf{A}\{\mathbf{x}(\boldsymbol{\lambda}, w)\}\mathbf{Q}, \tag{5.64}$$

where $\mathbf{Q}$ is the $n \times m$ matrix whose j-th row is $\nabla_{\boldsymbol{\lambda}} x_j(\boldsymbol{\lambda}, w)$.

Differentiating (5.60) with respect to $\boldsymbol{\lambda}$ gives

$$\hat{\mathbf{F}}\{\mathbf{x}(\boldsymbol{\lambda}, w), \boldsymbol{\lambda}, w\}\mathbf{Q} - \mathbf{A}^{\mathrm{T}}\{\mathbf{x}(\boldsymbol{\lambda}, w)\} = 0. \tag{5.65}$$

Substituting from (5.65) in (5.64) for $\mathbf{Q}$ gives the result (5.63).

(iii) From (i) and (5.59), ϕ is locally stationary at $\boldsymbol{\lambda}^*$. Since $\hat{\mathbf{F}}$ is positive definite at $\mathbf{x}^*$, $\boldsymbol{\lambda}^*$ and $\mathbf{A}(\mathbf{x}^*)$ is of full rank, $\nabla^2_{\boldsymbol{\lambda}}\phi$ from (5.63) is negative definite. Hence ϕ has an unconstrained maximum at $\boldsymbol{\lambda}^*$, and

$$\phi(\boldsymbol{\lambda}, w) \leqslant \phi(\boldsymbol{\lambda}^*, w) = f(\mathbf{x}^*) \quad \text{(from (5.61) (5.59))}.$$

Note that if conditions on f, $\mathbf{c}$ are such that $\mathbf{x}^*$ is a global constrained minimum for f, then $\boldsymbol{\lambda}^*$ is a global maximum for ϕ.

(6) If $\nabla_{\boldsymbol{\lambda}}\phi$ and $\nabla^2_{\boldsymbol{\lambda}}\phi$ are known, then Newton's method can be used to solve the equation

$$\nabla_{\boldsymbol{\lambda}}\phi = 0$$

whose root is $\boldsymbol{\lambda}^*$, by the iterative process

$$\boldsymbol{\lambda}_{r+1} = \boldsymbol{\lambda}_r - \{\nabla^2_{\boldsymbol{\lambda}}\phi\}^{-1}\nabla_{\boldsymbol{\lambda}}\phi = \boldsymbol{\lambda}_r - [\mathbf{A}\hat{\mathbf{F}}^{-1}\mathbf{A}^{\mathrm{T}}]^{-1}\mathbf{c}\{\mathbf{x}(\boldsymbol{\lambda}_r, w)\}. \quad (5.66)$$

(7) For very large w, the behaviour of $\hat{\mathbf{F}}$ is mainly controlled by the term $\frac{1}{2}w\sum_{i=1}^{m} c_i^2$ in f, and then it can be shown that

$$\{\nabla^2_{\boldsymbol{\lambda}}\phi\}^{-1} \simeq -w\mathbf{I}. \quad (5.67)$$

A multiplier algorithm for equality constraints The properties (1)–(7) of the last section suggest the design of an algorithm for the equality-constrained problem. Starting from the current point $\mathbf{x}_{r-1}$, current weighting w, and updated Lagrange multiplier $\boldsymbol{\lambda}_r$, the procedure is as follows.

Step 1 Minimise $f(\mathbf{x}, \boldsymbol{\lambda}_r, w)$ by an unconstrained-minimisation algorithm starting from $\mathbf{x}_{r-1}$ to find $\mathbf{x}_r$.

Step 2 Evaluate $\|\mathbf{c}(\mathbf{x}_r)\|$ as a measure of how well the constraints are satisfied. If this is zero, the constrained minimum is reached. If not, test the convergence of the process by comparing $\|\mathbf{c}(\mathbf{x}_r)\|$ with $\|\mathbf{c}(\mathbf{x}_{r-1})\|$. If the rate is too slow, e.g. if the ratio is >0.25, increase w to, say, $10w$ and repeat Step 1. If the convergence is satisfactory, go to Step 3.

Step 3 Update $\boldsymbol{\lambda}_r$ by using (5.66) or (5.67). Note that if the minimisation in Step 1 used a quasi Newton method, then an estimate of $\hat{\mathbf{F}}^{-1}$ is available for use in (5.66). Return to Step 1 with this $\boldsymbol{\lambda}_{r+1}$.

The process is started with an arbitrary w and $\boldsymbol{\lambda}$. The role of w is simply to ensure the correct behaviour of $\hat{f}$ and it will be kept constant once a suitable value is reached. By the result (5), the value of $\hat{f}(\mathbf{x}_r, \boldsymbol{\lambda}_r, w)$ is a lower bound on $f(\mathbf{x}^*)$.

A simple example is now given to demonstrate the working of the process.

Example 5.9 Minimise

$$f(\mathbf{x}) = 2x_1^2 + 2x_2^2 - 5x_1x_2 - 2x_1 - 11x_2$$

subject to $3x_2 - x_2 = 1$.
Note that the Hessian of f is not positive definite. The augmented objective is

$$\hat{f}(\mathbf{x}, \lambda, w) = 2x_1^2 + 2x_2^2 - 5x_1x_2 - 2x_1 - 11x_2 - \lambda(3x_1 - x_2 - 1) + \tfrac{1}{2}w(3x_1 - x_2 - 1)^2$$

The Hessian of this is

$$\hat{\mathbf{F}} = \begin{bmatrix} 4+9w & -5-3w \\ -5-3w & 4+w \end{bmatrix},$$

which is positive definite if $4+9w>0$ and $\det \hat{\mathbf{F}}>0$, i.e. $10w-9>0$. The unconstrained minimum of $\hat{f}$ is found to be

$$x_1(\lambda, w) = (63 + 42w + 7\lambda)/(10w - 9)$$
$$x_2(\lambda, w) = (54 + 116w + 11\lambda)/(10w - 9)$$

and the value of $c\{\mathbf{x}(\lambda, w)\}$ is $(144 + 10\lambda)/(10w - 9)$. With $\lambda = 0$ and $w = 1$, $x_1 = 105$, $x_2 = 170$, $c = 144$. A larger w, say $w = 10$, gives $x_1 = 5.31$, $x_2 = 13.34$, $c = 1.58$. The value of λ can now be updated, using (5.65) – which is exact for this case, since f is quadratic; the terms are

$$\nabla_\lambda \phi = -c\{\mathbf{x}(\lambda, w)\} = -1.58, \quad \nabla_\lambda^2 \phi = -0.11,$$

and so

$$\lambda_1 = 0 - (\nabla_\lambda^2 \phi)^{-1} \nabla_\lambda \phi = -14.4$$

The approximation (5.67) would give

$$\lambda_1 = -15.8$$

Repeating the first step with $\lambda = -14.4$ produces the solution

$$x_1^* = 4.2, \quad x_2^* = 11.6, \quad \text{and} \quad c = 0$$

so that this is the constrained minimum: by substitution, $f^* = -75.2$.

The dual function can be written down explicitly in this case; it is

$$\phi(\lambda) = -5(\lambda + 14.4)^2/(10w - 9) - 75.2$$

and it is clear that $\lambda^* = -14.4$ maximises this, and that the maximum value is

$$\phi^* = -75.2 = f^*.$$

Inequality constraints The same idea of augmenting the objective function can be used with inequality constraints. This is a developing subject, so mention will be made only of one idea, that due to Rockafeller (1973). To minimise $f(\mathbf{x})$ subject to $\mathbf{c}(\mathbf{x}) \geq 0$, the suggestion is to use

$$\hat{f}(\mathbf{x}, \boldsymbol{\lambda}, w) = f(\mathbf{x}) + \sum_i q_i(\mathbf{x}, \lambda_i, w)$$

where

$$\begin{aligned} q_i(\mathbf{x}, \lambda, w) &= -\lambda_i c_i(\mathbf{x}) + \tfrac{1}{2} w c_i^2(\mathbf{x}) \quad &\text{if } c_i(\mathbf{x}) \leq \lambda_i / w \\ &= -\lambda_i^2 / 2w &\text{if } c_i(\mathbf{x}) \geq \lambda_i / w \end{aligned}$$

with $w > 0$ as before. This produces an $\hat{f}$ which is continuous, and which behaves like f away from a constraint boundary, but like the function of the previous section near a boundary. Since for an active constraint at the minimum point $\mathbf{x}^*$,

$$c_i(\mathbf{x}^*) = 0, \quad \lambda_i^* \geq 0, \quad \text{then } q_i(\mathbf{x}^*, \lambda_i^*, w) = 0$$

and for an inactive constraint, $c_i(\mathbf{x}^*) > 0$, $\lambda_i^* = 0$, so that again $q_i(\mathbf{x}^*, \lambda_i^*, w) = 0$. Hence

$$f(\mathbf{x}^*, \boldsymbol{\lambda}^*, w) = f(\mathbf{x}^*).$$

The same properties can be shown to hold for this form, and the calculation can be carried through in the same way by alternately minimising to find $\mathbf{x}(\boldsymbol{\lambda}, w)$ and then updating $\boldsymbol{\lambda}$.
Example 5.10 Minimise

$$f(\mathbf{x}) = x_1^2 + x_2^2 + x_1 x_2 - 5x_1 - 7x_2$$

subject to $x_1 - x_2 \geq 1$.
Step 1 Assign $\lambda = 0$. Then

$$\hat{f}(\mathbf{x}) = \begin{cases} f(\mathbf{x}) \text{ in } c(\mathbf{x}) = x_1 - x_2 - 1 \geq 0, \text{ the feasible region,} \\ f(\mathbf{x}) + \tfrac{1}{2} w (x_1 - x_2 - 1)^2 \text{ elsewhere.} \end{cases}$$

Step 2 Find the unconstrained minimum of $\hat{f}$. This is at

$$x_1 = (1 + 5w)/(1 + 2w), \quad x_2 = (3 + 3w)/(1 + 2w)$$

and so with $w = 10$, the maximum is at $x_1 = 2.429$, $x_2 = 1.571$, and $c = -0.1428$. The point is outside the feasible region.

Step 3 Update λ. As before,

$$\nabla_\lambda \phi = -c(\mathbf{x}) = 0.1428, \quad \nabla_\lambda^2 \phi = -0.0952,$$

and so

$$\lambda = 0 + 0.1428/0.0952 = 1.5.$$

Now return to Step 1, with this λ:

$$\hat{f}(\mathbf{x}) = \begin{cases} f(\mathbf{x}) \text{ in } c(\mathbf{x}) \geq 0.15 \\ f(\mathbf{x}) - 1.5(x_1 - x_2 - 1) + \frac{1}{2}w(x_1 - x_2 - 1)^2 \text{ elsewhere.} \end{cases}$$

Repeating Step 2, this $\hat{f}$ has an unconstrained minimum at $x_1 = 2.5$, $x_2 = 1.5$, and c at this point is zero, so that this is the constrained minimum.

5.6 Summary of methods of constrained optimisation

The methods discussed in this chapter are projection methods (5.2), including gradient projection, reduced gradient, and Newton projection; quadratic programming methods (5.3), based on the simplex method or on projection; penalty and barrier function methods (5.5), where the constrained problem is replaced by a sequence of unconstrained problems using a parameter which tends to zero; and multiplier, or augmented Lagrangian, methods, where the unconstrained problems use estimates of the Lagrange multipliers which are modified as the calculation proceeds.

Problems

(1) Find the minimum of

$$f(\mathbf{x}) = 2x_1^2 - 2x_1x_2 + 2x_2^2 - 6x_1$$

subject to

$$-3x_1 - 4x_2 + 6 \geq 0, \quad x_1 - 4x_2 + 2 \geq 0,$$

with initial point $\mathbf{x}_0^T = (2, 0)$. Use

(a) the method of gradient projection,

(b) the reduced gradient method, taking $y_1 = -3x_1 - 4x_2 + 6$, $y_2 = x_2$,

(c) same method as (b), but using $y_1 = -3x_1 - 4x_2 + 6$, $y_2 = x_1 + x_2$.

(2) Find an initial feasible point for Problem 1, starting at (0, 1) and using the projection method of Section 5.2.2.

(3) Prove that if a nonsingular matrix $\mathbf{M}$ and its inverse $\mathbf{M}^{-1}$ are similarly partitioned into

$$\mathbf{M}=\begin{bmatrix}\mathbf{a} & \mathbf{b}\\ \mathbf{c} & \mathbf{d}\end{bmatrix}, \quad \mathbf{M}^{-1}=\begin{bmatrix}\mathbf{A} & \mathbf{B}\\ \mathbf{C} & \mathbf{D}\end{bmatrix}$$

where $\mathbf{d}$ is square and nonsingular, then

$$\mathbf{A}=(\mathbf{a}-\mathbf{b}\mathbf{d}^{-1}\mathbf{c})^{-1}, \quad \mathbf{B}=-\mathbf{A}\mathbf{b}\mathbf{d}^{-1}, \quad \mathbf{C}=-\mathbf{d}^{-1}\mathbf{c}\mathbf{A}, \quad \mathbf{D}=\mathbf{d}^{-1}(\mathbf{I}-\mathbf{c}\mathbf{B}).$$

Hence prove the results (5.33), (5.34).

(4) Gradient projection can be used to find an initial feasible point for the linear programming problem

$$\text{minimise } \mathbf{g}^{\mathrm{T}}\mathbf{x} \text{ subject to } \mathbf{A}\mathbf{x}=\mathbf{b}, \mathbf{x}\geqslant 0.$$

How is this related to the Phase 1 procedure of the Simplex method?

(5) The gradient projection method can be used to solve the linear problem stated in Problem 4, starting at an initial feasible point. How are the points produced related to those generated by the Simplex method?

(6) Let $c(\mathbf{x})\geqslant 0$ be a nonlinear constraint and $\mathbf{x}_0$ a point such that $c(\mathbf{x}_0)<0$, $|c(\mathbf{x}_0)|$ small. A first approximation to the direction of shortest distance from $\mathbf{x}_0$ to the feasible region is $\mathbf{a}(\mathbf{x}_0)$, where $\mathbf{a}(\mathbf{x})=\nabla c(\mathbf{x})$. Show that a better estimate of this direction is

$$\mathbf{p}=\{\mathbf{I}-c(\mathbf{x}_0)\mathbf{C}(\mathbf{x}_0)/\|\mathbf{a}(\mathbf{x}_0)\|^2\}\mathbf{a}(\mathbf{x}_0),$$

where $\mathbf{C}(\mathbf{x})$ is the Hessian of $c(\mathbf{x})$.

(7) Apply (a) the method of feasible directions, (b) the gradient projection method to the problem

minimise $\quad -3x_1-x_2$

subject to $\quad (x_1-2)^2+(x_2-1)^2\leqslant 9, \quad x_1, x_2\geqslant 0,$

starting at $\mathbf{x}_0^{\mathrm{T}}=(1, 1)$.

(8) Given that $\mathbf{A}_t$ is a $t\times n$ matrix and $\mathbf{A}_{t+1}$ is $\mathbf{A}_t$ augmented with a $(t+1)$st row $\mathbf{a}^{\mathrm{T}}$, obtain an expression for $\mathbf{A}_{t+1}\mathbf{G}^{-1}\mathbf{A}_{t+1}$ in terms of $\mathbf{A}_t\mathbf{G}^{-1}\mathbf{A}_t^{\mathrm{T}}$ and $\mathbf{A}_t$, $\mathbf{a}$, and $\mathbf{G}^{-1}$.

Prove that if

$$\mathbf{p}_{t+1} = (\mathbf{A}_{t+1}\mathbf{G}^{-1}\mathbf{A}_{t+1}^{\mathrm{T}})^{-1}\mathbf{A}_{t+1}\mathbf{G}^{-1}\mathbf{g}$$

$\mathbf{u}_{t+1}$ is the $(t+1)$st column of $(\mathbf{A}_{t+1}\mathbf{G}^{-1}\mathbf{A}_{t+1}^{\mathrm{T}})^{-1}$

$$\mathbf{p}_t = (\mathbf{A}_t\mathbf{G}^{-1}\mathbf{A}_t^{\mathrm{T}})^{-1}\mathbf{A}_t\mathbf{G}^{-1}\mathbf{g},$$

then

$$\begin{bmatrix} \mathbf{p}_t \\ 0 \end{bmatrix} - \mathbf{p}_{t+1} = -\frac{p_{t+1,t+1}}{u_{t+1,t+1}}\mathbf{u}_{t+1}.$$

(9) Verify that the projected gradient $\mathbf{p} = \mathbf{P}_0(-\mathbf{g})$ is the vector which minimises $(\mathbf{p}+\mathbf{g})^{\mathrm{T}}(\mathbf{p}+\mathbf{g})$, subject to $\mathbf{Ap} = 0$.

(10) Show that if the projected gradient method is used to minimise

$$f(\mathbf{x}) = \tfrac{4}{3}(x_1^2 - x_1x_2 + x_2^2)^{\frac{3}{4}} + x_3$$

subject to $\mathbf{x} \geqslant 0$, starting at $\mathbf{x}_0^{\mathrm{T}} = (a, 0, b)$, the process zigzags and does not converge to the true constrained minimum $\mathbf{x}^{*\mathrm{T}} = (0, 0, 0)$.

(11) The method of approximate programming for nonlinear objectives and constraints uses linear expansions about the current point and applies linear programming. Show that the general problem

$$\text{minimise } f(\mathbf{x}) \text{ subject to } \mathbf{c}(\mathbf{x}) \geqslant 0, \mathbf{x} \geqslant 0,$$

becomes, at the (feasible) point $\mathbf{x}_r$,

$$\text{minimise } \mathbf{g}(\mathbf{x}_r)^{\mathrm{T}}\mathbf{h}$$

subject to $\mathbf{x}_r + \mathbf{h} \geqslant 0$, $c_i(\mathbf{x}_r) + \mathbf{h}^{\mathrm{T}}\mathbf{a}_i(\mathbf{x}_r) \geqslant 0$, and $\mathbf{h}^{\mathrm{T}}\mathbf{h} \leqslant 1$. Test this method on Problem 7.

(12) Show that the solution $\mathbf{x}^*$ of the quadratic programming problem

$$\text{minimise } f(\mathbf{x}) = \tfrac{1}{2}x^{\mathrm{T}}\mathbf{G}\mathbf{x} + \mathbf{b}^{\mathrm{T}}\mathbf{x}$$

subject to $\mathbf{x} \geqslant 0$, where $\mathbf{G}$ is positive definite, satisfies

$$\mathbf{G}\mathbf{x}^* + \mathbf{b} \geqslant 0, \quad \text{and} \quad f(\mathbf{x}^*) = -\tfrac{1}{2}\mathbf{x}^{*\mathrm{T}}\mathbf{G}\mathbf{x}^*.$$

(13) Given the problem of maximising the total surface area of a circular cylinder of fixed volume V, show that the Hessian of the Lagrangian is indefinite. Verify that the augmented Lagrangian has an unconstrained maximum if $w > 1/(3\pi cV)$, where $c = (V/2\pi)^{\frac{1}{3}}$.

(14) Verify that if $\mathbf{x}^*$ is the solution of the problem

$$\text{minimise } f(\mathbf{x}) \text{ subject to } c(\mathbf{x}) \geqslant 0,$$

with f convex and c concave, and f, c differentiable, and if $c(\mathbf{x}^*) = 0$, then the minima $\mathbf{x}(r_k)$ of the penalty function

$$\hat{f}(\mathbf{x}) = f(\mathbf{x}) + r_k/c(\mathbf{x})$$

satisfy $\|\mathbf{x}(r_k) - \mathbf{x}^*\| \sim r_k^{1/2}$ as $k \to \infty$.

(15) Evaluate $\mathbf{x}(r_k)$ for the problem

$$\text{minimise } f(\mathbf{x}) = 4x_1 - x_2^2 - 12 \quad \text{subject to } x_1^2 + x_2^2 \leqslant 9,$$

using the penalty function of Problem 14, and $r = 1, \frac{1}{2}, \frac{1}{4}$. Assuming the behaviour of $\mathbf{x}(r_k)$ given in Problem 14, establish the acceleration procedure

$$\mathbf{x}^* \simeq \frac{\mathbf{x}_1 + (a + a^{\frac{1}{2}})\mathbf{x}_2 + a^{\frac{3}{2}}\mathbf{x}_3}{(a-1)(a^{\frac{1}{2}} - 1)},$$

where $\mathbf{x}_1 = \mathbf{x}(r)$, $\mathbf{x}_2 = \mathbf{x}(r/a)$, $\mathbf{x}_3 = \mathbf{x}(r/a^2)$, and test it on your results.

(16) Given the problem

$$\text{minimise } f(\mathbf{x}) = -x_1 + (x_1 + 2x_2)^2 \quad \text{subject to } x_2 - 2x_1 = 1,$$

define the augmented Lagrangian as

$$\hat{f}(\mathbf{x}, \lambda, w) = -x_1 + (x_1 + 2x_2)^2 - \lambda(x_2 - 2x_1 - 1) + \tfrac{1}{2}w(x_2 - 2x_1 - 1)^2.$$

Show that the dual function

$$\begin{aligned}\phi(\lambda) &= \hat{f}(\mathbf{x}(\lambda, w), \lambda, w)\\ &= (-0.08 + 0.4\lambda - 0.5\lambda^2 + 0.39w)/w.\end{aligned}$$

For what values of w has $\phi(\lambda)$ a maximum?

(17) A monopolist produces jointly two goods in quantities x_1 and x_2. The total cost of this production is $m(x_1, x_2) = x_1 + 2x_2 + 1300$. The relation between sales x_1, x_2 and prices p_1, p_2 per unit quantity are known to be

$$x_1 = 600p_2 - 400p_1, \quad x_2 = 1800 - 100p_1 - 300p_2.$$

What restrictions have to be imposed on the values of $\mathbf{x}$, $\mathbf{p}$? Find the values of $\mathbf{x}$, $\mathbf{p}$ subject to these restrictions which give maximum profit (i.e. maximum difference between return and cost).

(18) A firm is planning to make n different products; unit quantity of product j requires a_{ij} of material i, and they have b_i of this available. The cost of making unit quantity of product i is c_i and is constant; the selling price is a linear function of the x_i, of the form $s_i - \sum_j t_{ij}x_j$. Show that the determination of the production which gives maximum profit is a quadratic programming problem.

(19) A fund manager is choosing a portfolio of stocks in which to invest. The return from unit investment in stock j has been on average m_j and its variance has been s_j^2; the covariance of returns from stocks j, k is s_{jk}. If $\mathbf{S}$ is the variance–covariance matrix s_{jk}, with $s_{jj} = s_j^2$, then $\mathbf{S}$ is positive definite – since $s_{jk}^2 \leqslant s_j^2 s_k^2$ for all j, k – and the return from an investment of x_j in stock j, $j = 1, 2, \ldots, n$, has expected value $m = \sum_j m_j x_j$ and variance $s^2 = \mathbf{x}^{\mathrm{T}}\mathbf{S}\mathbf{x}$ (see any statistics text for these results). The fund manager's problem is to determine $\mathbf{x}$, satisfying $\mathbf{x} \geqslant 0, \sum x_j = 1$, such that the variance s^2 is least for given return m, or alternatively the expected return m is greatest for given s^2. Show that such $\mathbf{x}$ can be found by solving the quadratic programming problem

$$\text{maximise } f(\mathbf{x}, K) = K \sum_j m_j x_j - \mathbf{x}^{\mathrm{T}}\mathbf{S}\mathbf{x} \quad \text{for all } K > 0,$$

$$\text{subject to } \mathbf{x} \geqslant 0, \sum x_j = 1.$$

(20) Determine the best values of m, s^2 and $\mathbf{x}$ for the problem stated in 19, given that the choice is of two stocks only, with $m_1 = 0.15$, $s_1^2 = 0.0625$, $m_2 = 0.1$, $s_2^2 = 0.0225$, and $s_{12} = 0.0075$.

(21) Given N sets of readings $(x_{1k}, x_{2k}, \ldots, x_{nk}, y_k)$, $k = (1, 2, \ldots, N$, it is required to find the coefficients $c_1, c_2, \ldots, c_n$ which minimise the sum of squares

$$S = \sum_{k=1}^{N} (y_k - c_1 x_{1k} - c_2 x_{2k} - \cdots - c_n x_{nk})^2,$$

where $N > n$. Is S convex? Show that if there are inequality constraints on the c_j – e.g. if we require $c_j \geqslant 0$, all j – the problem is a quadratic programming one.

Hints and answers to problems

Chapter 1

(1) Stationary points at $(0, 0)$, $(\pm 1, 0)$. The Hessian is: at $(0, 0)$,

$$\begin{bmatrix} -1 & 0 \\ 0 & 1 \end{bmatrix}$$

indefinite, so this is a saddle point: at $(\pm 1, 0)$,

$$\begin{bmatrix} 2 & 0 \\ 0 & 3 \end{bmatrix}$$

positive definite, so these are minima.

(2) Stationary point $(6/5, 6/5, 17/5)$. Function is quadratic, with

$$\mathbf{G} = \begin{bmatrix} 4 & 1 & 0 \\ 1 & 2 & 1 \\ 0 & 1 & 2 \end{bmatrix}$$

positive definite, so this is the global minimum.

(3) (a) Obvious from definition. (b) $\mathbf{G}$ is the null matrix. (c) Requirement is that for all $\mathbf{h}$ such that $\mathbf{h}^T\mathbf{G}\mathbf{h} = 0$, the lowest nonvanishing term in the expansion of $f(\mathbf{x}^* + \mathbf{h})$ must be positive.

(4) Proved directly using the inequality $(a+b)^2 \geq 4ab > 3ab$ for $a, b > 0$.

(5) Both follow from definition of convexity.

(6) $\mathbf{x}^{*T} = (37/46, 16/46, 13/46)$, minimum; $\mathbf{g}(\mathbf{x}^*) = (7/23)\mathbf{a}_1(\mathbf{x}^*) + (2/23)\mathbf{a}_2(\mathbf{x}^*)$.

(7) Minimum.

(8) An example of the "abnormal case" – the constraint gradients are not independent.

(9)
$$\begin{aligned}
-3x_1^2 - 2x_3^2 - 2\lambda_1 - 10\lambda_2 x_1 - \mu_1 \qquad &= 0 \\
2x_2 \qquad -2\lambda_1 x_2 + 2\lambda_2 x_2 \quad -\mu_2 \qquad &= 0 \\
-4x_1 x_3 \qquad -\lambda_1 + \qquad \lambda_2 \qquad\qquad -\mu_3 &= 0
\end{aligned}$$

$\lambda_2(5x_1^2 - x_2^2 - x_3 - 2) = 0$, $\mu_1 x_1 = 0$, $\mu_2 x_2 = 0$, $\mu_3 x_3 = 0$, $\lambda_2, \mu_1, \mu_2, \mu_3 \geq 0$.

These conditions are satisfied at $\mathbf{x}^{*T} = (1, 0, 3)$ with $\lambda_1 = -47/4$, $\lambda_2 = 1/4$, $\mu_1\mu_1 = \mu_2 = \mu_3 = 0$.

(10) Minimum at $\mathbf{x}^{*T} = (10/9, 10/9, 40/9)$; c_2 active, $\lambda_2 = 20/9$, c_1 inactive.

(11) At maximum, $\mathbf{g} = \lambda\mathbf{p}$, $\mathbf{p}^T\mathbf{x} = d$, $\lambda > 0$, or $\mathbf{g} = 0$, where $\mathbf{g} = \nabla U$. Explicit if U is quadratic, so that $\mathbf{g} = \mathbf{Gx} - \mathbf{b}$. λ is the ratio between price and marginal utility, representing the rate at which maximum utility would change with increase in income.

(12) First part, direct from definition; second part, consider $\phi(\boldsymbol{\lambda}_1)$, $\phi(\boldsymbol{\lambda}_2)$ where $\boldsymbol{\lambda} = k\boldsymbol{\lambda}_1 + (1-k)\boldsymbol{\lambda}_2$, and prove that

$$k\phi(\boldsymbol{\lambda}_1) + (1-k)\phi(\boldsymbol{\lambda}_2) \leq \phi(\boldsymbol{\lambda}),$$

using

$$\phi(\boldsymbol{\lambda}_1) = \min_{\mathbf{x}\in X} f(\mathbf{x}) - \boldsymbol{\lambda}_1^T\mathbf{c}(\mathbf{x}) \leq f(\mathbf{x}) - \boldsymbol{\lambda}_1^T\mathbf{c}(\mathbf{x}) \quad \text{for any } \mathbf{x} \in X.$$

(13) Rayleigh's quotient. Result follows directly from orthogonality of eigenvectors.

(14) Direct from conditions at optimum.

(15) Direct from Kuhn–Tucker conditions.

(16) First use the Farkas' lemma to show that for any matrix $\mathbf{A} \neq 0$, and any matrix $\mathbf{B}$, then *either* $\mathbf{Ax} < 0$, $\mathbf{Bx} \leq 0$ has a solution *or* $\boldsymbol{\lambda}^T\mathbf{A} + \boldsymbol{\mu}^T\mathbf{B} = 0$, $\boldsymbol{\lambda} \geq 0$, with some $\lambda_i > 0$, $\boldsymbol{\mu} \geq 0$, has a solution but not both.

(17) As hint.

Chapter 2

(1) Minimum at $x = 2.83$, minimum value 42.26.

(2) Solve the difference equation to find I_1/I_N and compare with $(0.618)^N$.

(3) Cubic is $-3 - 4x + 5x^2 - 1.5x^3$ which has minimum value -3.939 at $x = 0.523$. The true minimum value of the function is -4 at $x = 0.5$.

(4) Points are (1, 1), (0.520, 1.096), (0.376, 0.376), (0.194, 0.412).

$$\text{Matrix} \quad \mathbf{P} = \frac{1}{\sqrt{5}}\begin{bmatrix} 1 & -2 \\ 2 & 1 \end{bmatrix}, \quad \lambda_1 = 1, \quad \lambda_2 = 11.$$

(5) If $r = 1$ the whole error is removed in one step, which only happens when $\mathbf{G}$ is the identity matrix.
Consider the product $(\mathbf{g}_K^T\mathbf{Q}\mathbf{g}_K)(\mathbf{g}_K^T\mathbf{Q}^{-1}\mathbf{g}_K)$, with $\mathbf{Q}$ diagonal, and show that it is decreased by the changes in the eigenvalues.

(6) Minimum at $(1, 2, -1)$.

(7) Stationary points at $(0, 3)$, saddle point, $(1, -0.5)$, minimum, and $(\pm 1, 0.5)$, saddle point.
Newton step from $(1, 1)$ approaches $(0, 3)$.

(8) Cycle 1 starts at $(0, 0, 0)$, value 0, directions $\mathbf{e}_1$, $\mathbf{e}_2$, $\mathbf{e}_3$.
Cycle 2 starts at $(2, 1, -0.6)$, value -9.4, same directions.
Cycle 3 starts at $(0.9, 2.1, -1.04)$, value -10.489, directions $\mathbf{e}_2$, $\mathbf{e}_3$, $\mathbf{s}$ where $\mathbf{s} = (-0.55, 0.55, -0.22)$, and finds minimum at $(1, 2, -1)$.

(9) From orthogonality of eigenvectors.

(10) Proof follows the same line as in Section 2.4.2.

(11) Minimum at $(0, 0, 0)$.

(12) Proof by induction.

(13) Immediate.

(14) Gives minimum in one step since the gradient $\mathbf{g}(\mathbf{x})$ is linear and so the secant formula is exact.

Chapter 3

(1) Solution $f^* = -16$, $x_1^* = 6$, $x_2^* = 4$. Other basic feasible solutions at $(0, 0)$, $(0, 6)$, $(4, 5)$, $(5, 2)$, $(3, 0)$.

(2) No feasible solution.

(3) No finite minimum.

(4) Consider the next variable to enter the basis; if the element in the same row as the zero is positive, it will be taken as the pivot, and the value f_k will be unchanged. By repeating the argument, show that, as long as the pivot is in the zero row, the variables which enter do not leave; and that ultimately a different pivot must become necessary, which will reduce f.

(5) Primal solution $x_1^* = 1/2$, $x_2^* = 1/2$; slack variables $x_3^* = x_4^* = x_5^* = 0$. This is a degenerate solution to the primal. The two dual solutions are

$$\lambda_1^* = 1, \lambda_2^* = 2, \lambda_3^* = 0; \quad \text{slack variables } \lambda_4^* = \lambda_5^* = 0$$

and

$$\lambda_1^* = 2, \lambda_2^* = 0, \lambda_3^* = 1; \quad \text{slack variables } \lambda_4^* = \lambda_5^* = 0,$$

and the general solution to the dual is the convex hull of these.

(6) Primal solution $x_1^* = 0$, $x_2^* = 2/3$, $x_3^* = 2/3$; $f^* = 4/3$.
Dual problem is

minimise $l = 2\lambda_1 + 2\lambda_2$

subject to $2\lambda_1 + 4\lambda_2 \geqslant 1$, $\lambda_1 + 2\lambda_2 \geqslant 1$, $2\lambda_1 + \lambda_2 \geqslant 1$,

with $\lambda_1, \lambda_2 \geqslant 0$. Solution is $\lambda_1^* = 1/3$, $\lambda_2^* = 1/3$, $l^* = 4/3$.

(7) Perturbations are multiples of the columns so involve no extra calculations.

(8) Immediate.

(9) Dual problem is

minimise $l=\lambda_1+4\lambda_2+3\lambda_3$

subject to $-\lambda_1+\lambda_2+\lambda_3\geqslant 3,\ \lambda_1+\lambda_2-3\lambda_3\geqslant 4,\ \lambda_1,\ \lambda_2,\ \lambda_3\geqslant 0.$

(10) Similar result to Problem 5.

(11) (a) Solution not optimal. New solution $(\frac{78}{25}, \frac{114}{25}, \frac{11}{10})$, $f^*=-583/50$.
(b) New solution $x_2^*=11/7$, $x_6^*=37/7$, $x_7^*=20/7$, $f^*=-93/7$.

(12) Because $\boldsymbol{\lambda}^*$ is feasible for new dual problem.

(13) Writing $x_s=x_s^{(+)}-x_s^{(-)}$, with $x_s^{(+)}$, $x_s^{(-)}\geqslant 0$, would produce the same final tableau with one extra column whose elements are those in the column for x_s with the signs reversed. Hence (a) if x_s is basic, the solution does not change; (b) if x_s is nonbasic, and if the elements in the x_s column are all nonnegative, then the new problem has an unbounded minimum; (c) if there is a negative element, then introducing $x_s^{(-)}$ into the basis will improve the solution. If x_s is a slack variable, the new problem corresponds to dropping the corresponding constraint.

(14) (a) Dual is

$$\text{maximise } p\lambda_1-10\lambda_2 \text{ subject to } \lambda_1-q\lambda_2\leqslant -1,\ \lambda_1-\lambda_2\leqslant -1,\ \lambda_1, \lambda_2\geqslant 0$$

(b) Unique solution if $q\geqslant 0$, $pq\leqslant 10$ or if $pq\geqslant 10$, $p\leqslant 10$.
(c) (i) Solution space empty if $p>10$, $pq>10$; dual then unbounded.
(ii) Solution space unbounded if $q<0$; dual then has no feasible solution.

(15) (a) Consider dual to problem

$$\text{maximise } \mathbf{g}^T\mathbf{x} \text{ subject to } \mathbf{A}\mathbf{x}=b, \mathbf{x}\geqslant 0.$$

(b) Immediate by considering dual slack and primal real variables.
(c) and (d) Special cases of (b).

Chapter 4

(1) A change in the level of an equality may or may not improve profit.

(2) Make 300 tonnes of P_1, none of P_2; 200 tonnes on M_1, 100 tonnes on M_2. Profit is £1540. Shadow prices £6.8 per hour on M_1, £0.6 per hour on M_2, zero for A and B.

(3) Determine possible partitions of the width W into the w_j and then use the lengths x_r to be cut according to each partition as variables in the LP.

(4) As hint, noting that if $x=P-Q$, and $P, Q\geqslant 0$, then $|x|\leqslant P+Q$, and so minimising $\sum(P_i+Q_i)$ minimises $\sum|x|$.

(5) Maximum output 48 750; 25, 75, 50 on systems 1, 2, 3. Dual variables give rate of change of output per change in the total numbers of men and of machines.

Maximum profit £4500 per week, with 150 on system 1. Dual variables give rate of change of profit with extra resources.

(6) (a) Minimise $\mathbf{a}_0^T\mathbf{x}$ subject to $(\mathbf{I}-\mathbf{A})\mathbf{x} \geqslant \mathbf{b}, \mathbf{x} \geqslant 0$, where all elements of $\mathbf{a}_0$, $\mathbf{A}$ and $\mathbf{b}$ are $\geqslant 0$.
(b) Solution only if $a_{ii} < 1$, all i, and if $b_j > 0$, all j, then $x_j > 0$, all j.
(c) $\mathbf{x}^* = (\mathbf{I}-\mathbf{A})^{-1}\mathbf{b}$.
(d) maximise $\mathbf{y}^T\mathbf{b}$ subject to $(\mathbf{I}-\mathbf{A})^T\mathbf{y} \leqslant \mathbf{a}_0, \mathbf{y} \geqslant 0$. All dual slack variables are zero, by complementary slackness, so $\mathbf{y}^* = \{(\mathbf{I}-\mathbf{A})^{-1}\}^T\mathbf{a}_0$, independent of $\mathbf{b}$.

(7) Multiple solutions, cost 211. $x_{12} = 7$, $x_{14} = 6$, $x_{21} = 5$, $x_{24} = 4$, $x_{32} = 3$, $x_{33} = 9$; $x_{13} = 7$, $x_{14} = 6$, $x_{21} = 5$, $x_{24} = 4$, $x_{32} = 10$, $x_{33} = 2$; $x_{12} = 3$, $x_{14} = 5$, $x_{21} = 5$, $x_{23} = 10$, $x_{32} = 7$, $x_{33} = 5$.

Add a dummy destination with costs zero. New solution also multiple, cost 209.

(8) $x_{11} = 10$, $x_{12} = 10$, $x_{23} = 20$, $x_{31} = 140$, $x_{33} = 40$, $x_{34} = 20$, cost 3850. Unsatisfied demand 50 at D_1.

(9) Sources are "bought new" or "from use on j-th day"; destinations are "for use on j-th day" or "left dirty after N-th day". Must use infinite cost for transit between "from j-th day" and "for k-th day" when $j \leqslant k$.

(10) New problem is transshipment; existing solution is feasible, with additional B. Check whether the dual of the new problem is feasible – if not, C^* can be reduced.

(11) Cost from year i to year j is c_{ij}, problem to minimise $\sum\sum x_{ij}c_{ij}$ where $x_{ij} = 1$ if machine is rented from i to j, otherwise $x_{ij} = 0$.

(12) Multiple solutions with costs 15.

(13) (a) Saddle point; strategies S_2, T_3, value 4.
(b) P_1's mixed strategy (0, 6/7, 0, 1/7); P_2's mixed strategy (9/14, 0, 0, 5/14); value 19/7.

(14) Blue; (1–1) split, proportion 3/5, (2–0) split, proportion 2/5.
Red: (2–1) split, proportion 1/5, (3–0) split, proportion 4/5.
In each case the larger force is allocated to A or B with equal chance. Value is 6/5 to Red. Note that optimal strategy for Red is convex hull of multiple solutions.

(15) By symmetry, play each strategy with proportion $1/m$.

(16) Use $\mathbf{y}^T\mathbf{A}\mathbf{x}^* \leqslant \mathbf{y}^{*T}\mathbf{A}\mathbf{x}^* \leqslant \mathbf{y}^{*T}\mathbf{A}\mathbf{x}$, and $\mathbf{y}^{*T}\mathbf{A}\mathbf{x}^* = \mathbf{v}$.

(17) Strategies "fire after i-th pace" with $i = 1, 2, \ldots, n$. $i = 0$ is dominated. For P_1 playing i and P_2 j, gain to P_1 is: $(i/n)(1-j/n) - j/n$ if $i > j$, is $(i/n) - (j/n)(1-i/n)$ if $i \leqslant j$. Mixed strategy for $n = 5$ is $i = 2$, 5/11; $i = 3$, 5/11; and $i = 5$, 1/11.

(18) Strategies for fighters are
(1) attack P first, attack P second unless P destroyed in first run.
(2) attack P first, attack F second.
(3) and (4) as (1) and (2) with P and F interchanged.

Strategies for bombers are (1) put V in P, (2) put V in F. The payoff to the fighters is

$F \backslash B$	1	2
1	$\gamma(2-\gamma)$	$\alpha\gamma$
2	γ	$\alpha\gamma+\beta(1-\gamma)$
3	$\alpha\beta$	$\beta(2-\beta)$
4	$\alpha\beta+\gamma(1-\beta)$	β

With values given for α, β, γ, F should play 4, $B1$, and value is 5/12.

(19) Banker's strategies:

	if Player stands	if Player draws 1	if Player draws 6
1	stand	stand	stand
2	stand	stand	draw
3	stand	draw	stand
4	stand	draw	draw
5	draw	stand	stand
6	draw	stand	draw
7	draw	draw	stand
8	draw	draw	draw.

Player's strategies:

1	stand with 1, stand with 6
2	stand with 1, draw with 6
3	draw with 1, stand with 6
4	draw with 1, draw with 6.

Banker's strategies 1, 4, 5, 8 are dominated, and 2, 6 and 3, 7 have the same payoff. Player should play 3 all the time; Banker any combination of 2 and 6; Banker gains 1/8.

(20) First result from convexity of f; second from convexity of feasible set.

(21) Use $y_1 = \ln x_1$.
Exact solution $x_1 = 3.57$, $x_2 = 1.25$ and $f = -0.55$.

Chapter 5

(1) $x_1 = 1.459$, $x_2 = 0.405$.

(2) $x_1 = 2/17 = 0.118$, $x_2 = 9/17 = 0.529$.

(3) Direct from $\mathbf{MM}^{-1} = \mathbf{I}$.

(4) Not directly.

(5) Same once an extreme point is reached.

(6) By estimating $\mathbf{a}(\mathbf{x})$ at the point on the constraint $c(\mathbf{x})=0$ for which the distance from $\mathbf{x}_0$ is least.

(7) $x_1=4.846$, $x_2=1.949$.

(8) Direct.

(9) From conditions on Lagrangian.

(10) Convergence is to $x_1=0$, $x_2=0$, $x_3=b-1.707a^{\frac{1}{2}}$. This example is due to Wolfe (1966).

(11) This method is due to Griffith and Stewart (1961).

(12) Direct from Kuhn–Tucker conditions.

(13) By the condition that the determinant of the Hessian of the augmented Lagrangian should be positive at the critical point.

(14) Using conditions at unconstrained minimum of f and at constrained minimum of f.

(15) This assumes an error $\|\mathbf{x}^*-\mathbf{x}(r)\|=Ar^{\frac{1}{2}}+Br$. Minimum is at $x_1=-2$, $x_2=2.236$.

(16) Direct from definition.

(17) Must have $\mathbf{x}\geqslant 0$, $\mathbf{p}\geqslant 0$. Optimum at $p_1=3.5$, $p_2=29/6=4.833$, $x_1=1500$, $x_2=0$.

(18) Immediate.

(19) A pair of values of m, s^2 are known as efficient.

(20) $x_1=0.214+0.357K$, $x_2=0.786-0.357K$. Corresponding values of m, s^2 can be found, for $K\leqslant 2.2$.

(21) S is convex – using the inequality $\{\sum(x_ix_j)\}^2\leqslant(\sum x_i^2)(\sum x_j^2)$.

References

Abadie, J. (ed.) (1967) *Nonlinear programming*, North-Holland.
Aoki, M. (1971) *Introduction to Optimisation Techniques*, Macmillan.
Beale, E. M. L. (1955) "Cycling in the Dual Simplex Algorithm", *Nav. Res. Log. Quart*, 269–75.
Broyden, C. G. (1972) "Quasi Newton methods" in Murray (1972), pp. 87–106.
Converse, A. O. (1970) *Optimisation*, Holt, Reinhart, and Winston.
Dantzig, G. B. (1963) *Linear Programming and Extensions*, Princeton University Press.
Davidon, W. (1959) "Variable metric methods for minimisation", *AEC R. and D. Report*, ANL-5990, Argonne Nat. Lab.
Dixon, L. C. W. (1972) "Quasi Newton methods generate identical points", *Math. Prog.*, **2,** 383–7.
Dixon L. C. W. and **Szego, G. P.** (eds.) (1975) *Towards Global Optimisation*, North-Holland.
Dresher M. (1961) *Games of Strategy — Theory and Applications*, Prentice-Hall.
Fiacco, A. V. and **McCormick, G. P.** (1968) *Nonlinear programming: Sequential Unconstrained Minimisation Techniques*, Wiley.
Fletcher R. (ed.) (1969) *Optimisation*, Academic Press.
Fletcher, R. (1971) "A general quadratic programming algorithm", *JIMA*, **7,** 76–91.
Fletcher, R. (1975) "An ideal penalty function for constrained optimisation", *JIMA*, **15,** 319–42.
Fletcher, R. and **Powell, M. J. D.** (1963) "A rapidly convergent descent method for minimisation", *Comp. J.*, **6,** 163–8.
Fletcher, R. and **Reeves, C. M.** (1964) "Function minimisation by conjugate gradients", *Comp. J.*, **7,** 149–54.
Ford, I. R. and **Fulkerson, D. R.** (1962) *Flows in Networks*, Princeton University Press.
Gass, S. I. (1958) *Linear Programming*, Addison-Wesley.
Gill, P. E. and **Murray, W.** (1972) "Quasi Newton methods for unconstrained optimisation", *JIMA*, **9,** 91–108.
Gill P. E. and **Murray, W.** (1973) "Quasi Newton methods for linearly constrained optimisation", *NPL Report* VAC 32.
Gill, P. E. and **Murray, W.** (eds.) (1974) *Numerical methods for constrained optimisation*, Academic Press.
Goldfarb, D. (1969) "Extension of Davidon's variable metric method to maximisation under linear inequality and equality constraints", *SIAM J. Appl. Math.*, **17,** 739–64.

Griffith, R. E. and **Stewart, R. A.** (1961) "A nonlinear programming technique for the optimisation of continuous processing systems", *Management Sci.*, **7,** 379–92.

Hadley, G. (1961) *Linear Programming,* Addison-Wesley.

Hadley, G. (1964) *Nonlinear and Dynamic Programming,* Addison-Wesley.

Hestenes, M. R. and **Stiefel, E.** (1952) "Methods of conjugate gradients for solving Linear Systems", *J. Res. Nat. Bur. Standards,* **49,** 409–36.

Hooke, R. and **Jeeves, T. A.** (1961) "Direct search solution of numerical and statistical problems", JACM, **8,** 212–29.

Huang, H. Y. (1970) "Unified approach to quadratically convergent algorithms for function minimisation", *JOTA*, **5,** 405–23.

Kuhn, H. W. and **Tucker, A. W.** (1951) "Nonlinear programming" in *Proceedings of the Second Berkeley Symposium on Mathematical Statistics and Probability* ed. Neyman, Univ. California Press, 481–93.

Levenberg, K. (1944) "A method for the solution of certain problems in least squares", *Q. Appl. Math.*, **2,** 164–8.

Lootsma, F. A. (ed.) (1972) *Numerical Methods for Nonlinear Optimisation,* Academic Press.

Luenberger, D. G. (1973) *Introduction to Linear and Nonlinear Programming,* Addison-Wesley.

McKinsey, J. (1952) *Introduction to the Theory of Games,* McGraw-Hill.

Mangasarian, O. L. (1969) *Nonlinear Programming,* McGraw-Hill.

Marquardt, D. (1963) "An algorithm for least squares estimation of non-linear parameters", *SIAM J. Appl. Math.*, **11,** 431–41.

Miller, C. E. (1963) "The simplex method for local separable programming", in *Recent Advances in Mathematical Programming,* ed. Graves, R. L. and Wolfe, P., McGraw-Hill.

Murray, W. (ed.) (1972) *Numerical Methods for Unconstrained Optimisation,* Academic Press.

Nelder, J. A. and **Mead, R.** (1965) "A simplex method for function minimisation", *Comp. J.*, **7,** 308–13.

Pierre, D. A. and **Lowe, M. J.** (1975) *Mathematical Programming via Augmented Lagrangians,* Addison-Wesley.

Polak, E. (1971) *Computational Methods in Optimisation,* Academic Press.

Powell, M. J. D. (1964) "An efficient method of finding the minimum of a function of several variables without calculating derivatives", *Comp. J.*, **7,** 155–62.

Rockafeller, R. T. (1973) "A dual approach to solving nonlinear programming problems by unconstrained optimisation", *Math. Prog.*, **5,** 354–73.

Rosen, J. B. (1960) "The gradient projection method for nonlinear programming: Part I, linear constraints", *SIAM J. Appl. Math.*, **8,** 181–217.

Rosen, J. B. (1961) "The gradient projection method for nonlinear programming: Part II, nonlinear constraints", *SIAM J. Appl. Math.*, **9,** 514–32.

Sargent, R. W. H. and **Murtagh, B. A.** (1973) "Projection methods for nonlinear programming", *Math. Prog.*, **4,** 245–68.

Swann, W. H. (1974) "Constrained optimisation by direct search", in Gill and Murray (1974).
Vajda, S. (1962) *Readings in Mathematical Programming*, Pitman.
Vajda, S. (1967) "Nonlinear programming and duality" in Abadie (1967).
Varaiya, P. P. (1972) *Notes on Optimisation*, Van Nostrand.
Wilde, D. and **Beightler, C. S.** (1967) *Foundations of Optimisation*, Prentice-Hall.
Wolfe, P. (1959) "The simplex method of quadratic programming", *Econometrica*, **27**, 382–98.
Wolfe, P. (1961) "A duality theory for nonlinear programming", *Q. J. Appl. Math.*, **19**, 239–44.
Wolfe, P. (1966) "On the convergence of gradient methods under constraints", *Res. Rept.*, IBM Zurich.
Wolfe, P. (1967) "Methods for nonlinear constraints", in Abadie (1967).
Zoutendijk, G. (1960) *The Method of Feasible Directions*, Elsevier.

Index

active constraint, 12, 133, 134, 162
assignment problem, 109–16
augmented Lagrangian, 158

barrier-function method, 151
basic feasible solution, 74
basic variable, 77
basis, 74

complementary slackness, 90
concave function, 19
conjugate
  directions, 44, 49
  gradients, 46–9, 62
constraint qualification, 18, 30
constraints
  active, 12, 133, 134, 162
  equality, 6–12, 71, 156
  inactive, 13, 133, 140
  inequality, 12–15, 133
  linear, 70, 134
  nonlinear, 147, 149, 151
convex
  cone, 15
  duality, 25
  function, 4, 19
  hull, 20
  set, 15, 20, 71
cubic fit, 40
cycling, 86, 95

Davidon–Fletcher–Powell method, 56–9
degeneracy, 77, 110
directional derivative, 2
dual
  function, 24
  problem, 24
  simplex method, 91
duality theorems
  convex, 25–26
  linear, 86–9
  symmetric, 27–8

equality constraint, 6–12, 71, 156
extreme point, 71
Euclidean norm, 1

Farkas lemma, 16–8
feasible
  direction, 12, 134, 149
  point, 141
  region, 12
  solution, 74
Fibonacci search, 35–7
first order conditions, 2, 6, 13

games, theory of, 116–22
global minimum, 1, 5, 19, 134
golden section, 37
gradient
  of function, 2, 14
  projected, 134
  reduced, 136

Hessian matrix, 2, 54, 64
Huang methods, 59–62
hyperplane, 71

implicit function theorem, 7, 30
inactive constraint, 13
inequality constraint, 12

Kantorovich inequality, 64
Konig's theorem, 111
Kuhn–Tucker conditions, 13, 133

Lagrange multipliers, 10, 90, 138
Lagrangian, 10, 90,
Least squares methods, 63
line search, 35–40
linear programming, 70

minimum
  local, 1, 6, 12, 13, 29, 134
  global, 1, 5, 19, 134

Newton's method
  for unconstrained minimisation, 54
  projected, 146

pattern search, 40
penalty function methods, 150–5
perturbation, 95, 110
phase 1, simplex method, 83
pivot element, 79
positive definite matrix, 3
prisoners' dilemma, 116
projected gradient, 134

quadratic
  fit, 38
  function, 3, 44–6, 55
  programming, 145–7
  termination, 44
quasi Newton methods, 54–69

rank one correction, 63
rank two correction, 59
reduced gradient, 136
regular point, 7
revised simplex method, 83–6

saddle point, 23, 90, 117
search
  line, 35
  pattern, 40
second-order conditions, 2, 11, 20
sensitivity, 9, 14, 91
separable programming, 122–5
sequential unconstrained minimisation (SUMT), 150
scaling, 42
shadow prices, 98
simplex method
  dual, 91
  primal, 77
  revised, 83
simplex tableau, 78
slack variable, 70
solution set, 12
standard form
  of linear problem, 70
  of nonlinear problem, 12
stationary point, 2
steepest descent, 42–3
strategy
  mixed, 118
  optimal, 119
  pure, 117
surplus variables, 81
symmetric duality theorem, 27

transportation, 100–6
transshipment, 106–9
two-person game, 116
two-phase simplex method, 81

unbounded solution set, 73
unsymmetric duality, 24

value, of a game, 118

zero sum game, 116